AVIATION MYSTERIES OF THE NORTH

DISAPPEARANCES IN ALASKA AND CANADA

Gregory P. Liefer

PO Box 221974 Anchorage, Alaska 99522-1974
books@publicationconsultants.com—www.publicationconsultants.com

ISBN 978-1-59433-195-4
eBook ISBN 978-1-59433-209-8
Library of Congress Catalog Card Number: 2011921682

Copyright 2011 Gregory P. Liefer
—First Printing 2011—
—Second Printing 2011—
—Third Printing 2015—

All rights reserved, including the right of reproduction in any form, or by any mechanical or electronic means including photocopying or recording, or by any information storage or retrieval system, in whole or in part in any form, and in any case not without the written permission of the author and publisher.

Manufactured in the United States of America.

For Nellie and Wilma

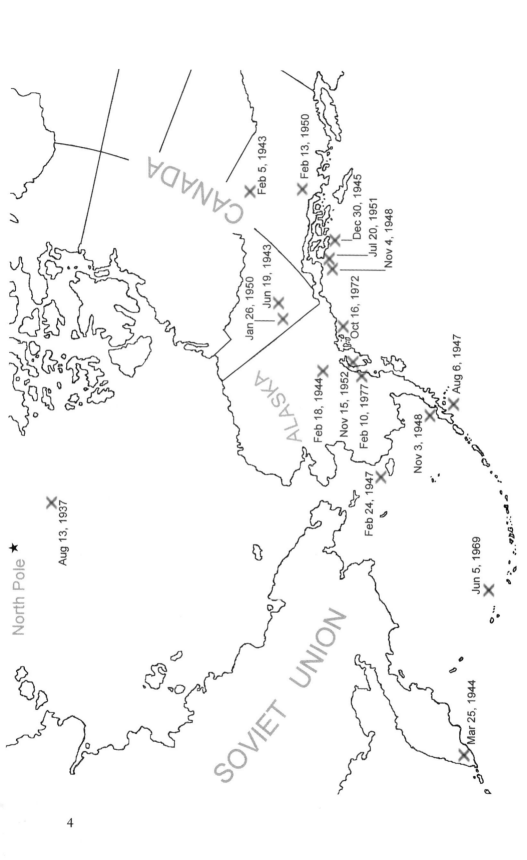

Acknowledgments

The historical events described in this book could not have been written without the assistance of numerous individuals and organizations. While much of the information was obtained from government agencies and academic institutions, the contributions from public archives and private collections are equally important. For the following people who provided their time and resources, and anyone else I may have overlooked, thank you.

Randy Acord, Alaska aviation pioneer, historian, and former Ladd Field test pilot, for thoughtful insight and access to his personal aviation collection; Author Carl Mills for research and comments on the missing Canadian Pacific Airlines DC-4; and author Blake Smith for suggestions and historical aircraft images of the Northwest Staging Route.

Lieutenant Colonel Gerald O'Hara and Lieutenant Ken Hall of the Joint POW/MIA Accounting Command (formerly the Central Identification Laboratory), Public Affairs Office; Sandy Williams and Phyllis Shaw at the Department of the Navy, Office of the Judge Advocate General; Scott Price and historian Chris Havern of the United States Coast Guard; Alley Louie at the Air Force Historical Research Agency; Barry Cottam and Tim Cook at the National Archives of Canada; and Nick Williams at the Transportation Safety Board of Canada; for declassified documents, investigation material and aircraft accident reports.

John Cloe and James Frank at the 3rd Wing Historical Office, Elmendorf AFB, Alaska; Technical Sergeant Donald Fenton at the 354th Wing Historical Office, Eielson AFB, Alaska; Katherine Williams at the Museum of Flight, Seattle, Washington; Laura Kissel at the Ohio State University Archives; Robin Weber at the Northwest Territories Archives; Ian Leslie and Fiona Smith Hale at the Canada Aviation Museum Archives; Steven Maxham at the US Army Aviation

Museum; Tim Travis at Raytheon Aircraft; Ray Owenby at Lockheed Martin Aircraft; King Hawes at 6srw.com; Ron Dupas at 1000aircraftphotos.com; Brian Lockett at Air-and-Space.com; Claus Martel at the US Army Aviation and Missile Command; Joe Stevens at the Kodiak Military History Museum; author A.T. Lloyd; aviation artists Jack Fellows and Paul Fretts; Air Classics magazine; Doug Davidge; Mike Hewitt; Daniel Jessup; Steve Hawley; Howard Wynia; Francis Boisseau; Chris Charland; Peter Bresciano; Earl Rogers; Chuck Lunsford; Danny Martin; Ed Steffen; Logan Delp; Robert "Bob" Brown; Jason Parsons; Earl Brown; Georgie Behn; Ed Coates; Dan Lange; Tony Suarez; George Villasenor; and Robert Koppen; for their photographic image contributions.

Last and most importantly, I thank my wife Liz and daughter Lauren for their loving support, without which this book would never have been written.

Contents

Dedication 3
Acknowledgments 5
Preface 9
Chapter One .... Across The North Pole 11
Chapter Two .... Secret Cargo 37
Chapter Three .... All Reports Normal 53
Chapter Four .... Caught in the Air 63
Chapter Five .... Empire Express 73
Chapter Six .... A Cold Winter Morning 91
Chapter Seven .... Weather Reconnaissance 105
Chapter Eight .... Edge of the Storm 117
Chapter Nine .... A Routine Training Flight 129
Chapter Ten .... Fading Signal 141
Chapter Eleven .... Without a Trace 153
Chapter Twelve .... Broken Arrow 169
Chapter Thirteen .... Cathedrals of Ice 189
Chapter Fourteen .... Operation Warm Wind 205
Chapter Fifteen .... Last Flight of Irene 92 217
Chapter Sixteen .... Accident or Conspiracy? 229
Chapter Seventeen .... Cleared As Filed 243
Bibliography 257
Index 261

Preface

Aviation Mysteries of the North: Disappearances in Alaska and Canada, is a descriptive account of lost aircraft for more than four decades of aviation history in one of the harshest and more remote regions on earth. Many of the disappearances received little or no publicity due to classified circumstances involving political and strategic interests, while others were overshadowed by aviation events occurring much closer to the major media outlets of the world. Although well known stories such as Amelia Earhart and the Bermuda Triangle continue captivating the public's interest, the incidents detailed in this book are equally entrenched in mystery and should never be forgotten.

In the early years of aviation flying was often performed without the use of accurate maps, using only basic flight instruments. As technology progressed with faster and larger aircraft, navigational aids were established, routes were charted and instruments improved, allowing flights in conditions previously unimagined. Weather and inhospitable terrain were no longer a deterrent, but a challenge. Man and machine began flying at all times, to all places, in all situations, eventually becoming a matter of routine.

During the transition of aviation from early barnstormers to modern day commercial airline travel, flights were much more dangerous and much less forgiving. Airways and cockpit instruments were not always reliable, training and certification standards were less stringent, and a capability for forecasting some of the most extreme flight hazards was unavailable. In the past and even today, technology remains dependent on the capability of the person using it. Even the most experienced individuals are not infallible. Whether it is the technology or the individual that ultimately fails, disastrous consequences can and do occur. As a result, sometimes aircraft disappear and are never found. More than a hundred aircraft, from small single-engine bush planes to large multi-engine military transports and commercial airliners have vanished along the cold waters and vast wilderness of the North. A few were found years or even decades later. Most remain missing today.

The seventeen stories detailed in *Aviation Mysteries of the North: Disappearances in Alaska and Canada* are drawn from government accident reports, declassified documents, newspaper articles, interviews, historical accounts and personal observations. I have attempted to explain the circumstances surrounding each incident in a less technical format that provides a human element to the mysteries, for these were not just machines, but individuals and families who never imagined the tragic fate ahead of them on the horizon.

Chapter One

August 13, 1937

Across The North Pole

Secrecy shrouded the takeoff of Soviet aviator Sigismund Levanevsky and his five-man crew aboard a four engine Bolkhovitinov DB-A aircraft from Moscow in August 1937. For weeks preceding the departure only small bits of information regarding the proposed flight across the North Pole were released to the world by Soviet dictator Joseph Stalin's communist regime. Many details were not revealed until absolutely necessary, and only then in the briefest of formats. The exact route and final destination after crossing the Pole were even kept secret until the day of departure.

Fairbanks, Alaska was announced early on as the first landing point, but from there the details of the planned flight across Canada into the United States were left to speculation. The flight was to be Russia's third attempt at a successful trans-polar crossing in as many months, with the exception this one would be commanded by Russia's most famous aviator, Sigismund Levanevsky, often referred to by the news media as the "Soviet Lindbergh." Unlike the two previous polar flights, the intent of Levanevsky was not to establish another non-stop world distance record, but rather to show the feasibility of a commercial air route between the Soviet Union and the United States.

A similar attempt by Levanevsky and two crewmen was made in August 1935, but the flight was terminated after nine hours when their single-engine Tupolev ANT-25 developed an oil leak. Undeterred by the failure, Levanevsky immediately began planning another attempt, even though Stalin directed the country's aviation resources be focused in a different direction. Developing new aviation capabilities that could enhance the world's view of the communist system became a new priority. During the 1930s the world non-stop distance flight record was an internationally publicized event in the aviation community and achieving a new record became the primary objective of the Soviet government.[1]

That goal seemed to be fulfilled in June and July 1937, when two separate Soviet aircraft established distance records flying non-stop from Russia to the United States over the North Pole. The aircraft were single-engine Tupolev ANT-25s, the same type Levanevsky had flown on his trans-polar attempt in 1935, but with new modifications. Both flights received great publicity and international acclaim, as well as being recognized as legitimate world records. The authenticity of the routes and distances flown by the Soviets during the flights were brought into question years later[2], but the details were validated by several investigative sources over the ensuing decades. Only after the two successful flights were completed did Stalin allow Levanevsky another attempt at his dream of establishing a polar air route between Asia and North America.

Sigismund Levanevsky was a relatively new and unknown aviator in the Soviet Union until 1933. He first gained attention in the summer of that year after rescuing American pilot Jimmy Mattern in Siberia, following the crash of Mattern's plane during an around-the-world speed record attempt. Mattern was listed as missing for weeks, sparking an international interest until he was found by a group of Eskimos near the Anadyr River in Eastern Siberia. Stalin immediately saw the potential for publicity and used the incident to great benefit. A seaplane flown by Levanevsky was quickly dispatched to pick-up the stranded American pilot and transport him across the Bering Strait to Nome, upon which Levanevsky received a hero's welcome and world-wide recognition. Soon after, Soviet news agencies hailed the young pilot's accomplishments on a par with Charles Lindbergh.

Almost a year after Mattern's timely rescue, Sigismund Levanevsky was involved in another incident which enhanced his fame as an aviator. When a Russian freighter exploring a winter route across the Arctic Ocean became trapped in sea ice off the North Cape of Siberia, an aerial rescue seemed the only possibility of saving the survivors. The ship *Cheliuskin* had originally left Murmansk in August 1933 before battling shifting ice flows for more than 3,000 miles on the long journey north around the Asian Continent. But by October the *Cheliuskin* was firmly trapped in the ice, carried along at the mercy of northern currents for months until the situation became hopeless. As the ship started splintering apart and sinking in February 1934, a distress call was finally broadcast. All the occu-

---

1 Between 1933 and 1938 the Soviet Union established 62 world flight records
2 In his book *Russia's Shortcut to Fame*, Robert J. Morrison claims the Soviet trans-polar flights in 1935 and 1937, including Levanevsky's, were giant deceptions perpetrated by Stalin to exaggerate the success of Soviet aviation.

pants, including scientists, crew and passengers were moved off the ship onto the icepack. Luckily the survivors had adequate time before the ship sank to move several months of food and supplies with them.

In spite of the seriousness of the situation, Stalin would not request assistance in the rescue of the *Cheliuskin* survivors, probably in a distorted belief that the Soviet Union would be discredited. Instead, he dispatched Levanevsky and Mavriki Slepnyov to the United States to purchase two aircraft for use in a Soviet rescue attempt. Both pilots arrived with typical media fanfare, especially Levanevsky, whose friendly demeanor and ability to speak English quickly made him a celebrity. Only after weeks traveling

Levanesvsky returning American pilot Jimmy Mattern to Nome, Alaska in the Dornier-Wal seaplane. Levanevsky is in the middle, shaking Mattern's hand. (Yuri Kaminsky Collection via Mike Hewitt)

across the United States were two relatively new Consolidated Fleetsters finally purchased from Pacific Alaska Airways in Fairbanks and flown to Nome. Slepnyov and Levanevsky, accompanied by American mechanics Clyde Armistead and Bill Lavery, left the coastal Alaskan settlement in the two aircraft on 24 March, flying west across the Bering Strait before turning north along the Siberian coast. Other Soviet aircraft from distant bases were by then nearing the northern coast of Siberia as well.

While Levanevsky was en route near North Cape, not far from the stranded survivors, he was forced to make an emergency landing, seriously damaging the plane.³ Unable to continue, he was eventually rescued while the *Cheliuskin* survivors were flown off the ice by other pilots during the ensuing weeks. Fortunately for Levanevsky, Stalin did not see his crash as a failure and awarded him and six other pilots involved in the rescue the "Order of the Hero of the Soviet Union,"

Levanevsky and Levchenko in the Vultee V-1A taxiing into the beach at Harding Lake, Alaska, near Fairbanks, on August 11, 1936. (Pioneer Aviation Museum-Randy Acord)

American mechanic Clyde Armistead greets Levchenko at Harding Lake while Levanevsky stands in the cockpit. (Pioneer Aviation Museum-Randy Acord)

the country's highest decoration. With his newly bestowed honor and the sensationalized media coverage he obtained while in the United States, Levanevsky's reputation became well established.

---

3 North Cape was later renamed Cape Schmidt in honor of the *Cheliuskin* expedition leader.

It was during the *Cheliuskin* rescue that Levanevsky first recognized the potential for a commercial air route across the North Pole. Using his new prestige and access to Stalin's inner circle, he began planning to fulfill that dream in the next few years.

Levanevsky returned to Alaska in August 1936 after purchasing a newly designed single-engine Vultee V-1A floatplane for the Soviet government in Cali-

Standing left to right: Levchenko and Levanevsky with American mechanics Armistead and Lavery, who assisted with the Cheliuskin rescue in 1934. (Pioneer Aviation Museum-Randy Acord)

fornia. Intending to record high altitude atmospheric conditions in the Arctic and test the feasibility of an air route between North America and Asia, Levanevsky was accompanied by fellow pilot and navigator, Victor Levchenko. They completed the 10,000 mile journey from California to Moscow in two weeks, setting the stage for further Soviet long-range flights.

Unfortunately for Levanevsky, the two trans-polar flights which followed were flown by other Soviet aviators in June and July 1937, leaving him and the international media pondering when the "Soviet Lindbergh" would be given his opportunity. It finally came a month later in August, when Levanevsky and a five man crew departed Moscow on their own flight attempt across the North Pole to North America. The crew consisted of co-pilot Nikolai Kastanayev, navigator Victor Levchenko, mechanics Victor Probezhimov and Nikolai Godovikov and radio operator Nicolai Galkovsky. Levchenko had previously accompanied Levanevsky on his first trans-polar attempt in 1935 and their high altitude flight through Alaska in 1936.

Local residents look over Levanevsky's Vultee V-1A at Harding Lake, Alaska. (Pioneer Aviation Museum-Randy Acord)

Radio and weather stations necessary for accurate information and communications were already in place over much of Levanevsky's proposed route across northern Siberia, Alaska and Canada by August 1937. A remote weather station near the North Pole had even been established by the Soviets months earlier and was manned by a small contingent of Russian scientists. Coordination was also made with Alaskan and Canadian weather stations along the route, allowing updated reports to be transmitted every three hours instead of the customary six. Because of the language difficulties and a lack of available translators at the American and Canadian stations, the Soviets placed a Russian radio operator in Seattle, Washington to translate radio messages for the press and another operator to assist with communications in Fairbanks.

Originally scheduled for departure from Moscow in late July, Soviet news releases repeatedly claimed weather conditions along the 4,200 mile route from Moscow to Fairbanks kept the aircraft grounded. Although it is certainly probable mechanical and logistical problems contributed to the delayed departure, the Soviet government only confirmed poor weather was a contributing factor.

Reports of unfavorable weather along the route became an almost daily occurrence, continuing into August. At the same time there were only hints of where the plane's final destination would be. At first San Francisco was reported as the destination, then New York and Chicago. Other details were released sporadically, keeping the world's interest at a peak.

The type of aircraft and its capabilities were also obvious news items that were only released piecemeal. Eventually it was learned the aircraft to be flown by Levanevsky and his crew on the trans-polar flight attempt was a newly designed four engine transport having a 129 foot wingspan and a gross weight of 72,000 pounds. The plane was designed by Russian engineer Viktor Bolkhovitinov and given the designation DB-A by the Soviet government.[4]

The Vultee V-1A being refueled at Harding Lake, Alaska. (Pioneer Aviation Museum-Randy Acord)

A prototype initially designed as a heavy bomber had first flown in 1935 and since been subjected to various test flights and modifications. In place of bombs, Levanevsky's plane carried extra fuel tanks needed for the long range flight over the Pole. Given a Soviet registration of H-209, its capabilities were listed as a maximum speed of 230 mph, a ceiling of 23,000 feet and a range of 8,000 miles on 7,000 gallons fuel capacity. Each of the M-34RN engines was rated at 850 hp and recessed back inside thick wings for internal access during flight by two on-board mechanics. The aircraft was similar in size to the older Soviet Tupolev TB-3

---

4 Levanevsky's plane has often been misidentified in various publications as a Tupolev TB-3, ANT-4 or ANT-6.

bomber, but was much more aerodynamic and fuel efficient with a smooth metal fuselage and enclosed cockpit. The main landing gear could also be retracted into large metal housings hanging beneath the wings, enclosing the engine radiators.

Levanevsky claimed the aircraft could easily fly on three engines if necessary, although the maximum ceiling would be reduced to 15,000 feet. He stated that icing was not anticipated as a problem, since their intent was to climb above any cloud formations during the flight. Conditions inside the aircraft were reported as comfortable compared to other Soviet aircraft, with temperatures consistently in the fifties even when the outside temperature dipped well below zero. Survival equipment onboard included a rubber raft, rifles, tools, oxygen canisters, six weeks supply of food and plenty of warm clothing.

On the evening of August 12, 1937, Levanevsky and his crew finally departed the Shelkova military airfield outside Moscow.[5] Their plan after passing over the North Pole was to continue south, following along the 148 degree meridian into Fairbanks, Alaska. Only after takeoff was it finally announced the flight was expected to take approximately thirty hours to Fairbanks, where the plane would be refueled before continuing on with subsequent stops at Edmonton, Alberta, Chicago and New York.

Sigismund Levanevsky at Sevastpol in May 1937. (Yuri Kaminsky Collection via Mike Hewitt)

A Soviet DB-A transport with its distinctive fuselage and landing gear. (1000aircraftphotos.com)

---

5 Fairbanks, Alaska time was 5:15 AM, thirteen hours behind the Moscow departure time of 6:15 PM

Apparently the Soviets had tired of waiting on perfect weather for the flight, since two large weather systems over the Arctic Ocean and a smaller system over the Siberian Sea were reported moving in the direction of the aircraft's intended route. All three weather systems were accompanied by high winds and significant cloud accumulations.

Levanevsky's DB-A transport, registration number H-209, before departure from Shelkova airfield outside Moscow. (Yuri Kaminsky Collection via Mike Hewitt)

Levanevsky poses with his five man crew shortly before departure. (Yuri Kaminsky Collection via Mike Hewitt)

The community of Fairbanks, Alaska, which had been preparing for the flight's anticipated arrival for days, finished the final arrangements after receiving confirmation the plane finally departed. Since the large plane was expected to land at Weeks Field on the edge of town around noon the following day, the runway was kept clear of unnecessary traffic and an elaborate reception planned for the Soviet crew that evening. Following the request of the Soviet government, 3,000 gallons of fuel and other supplies had already been set aside for the plane's arrival. It would be the first aircraft of that size to land at Fairbanks.

The flight seemed to proceed as planned following the takeoff from Moscow. A short transmission checking the operation of the radio and confirming all aircraft systems were normal was sent by the crew ten minutes after departure, then a similar transmission again twenty minutes later. Routine position reports followed as Levanevsky and his crew proceeded north toward Archangel on the Barents Sea. At 8:30 am Fairbanks time, a little more than three hours after takeoff, the plane was still south of Archangel, but continuing as scheduled. Additional radio transmissions were received by other Soviet stations at Amderma

H-209 departing Shelkova airfield on the trans-polar flight attempt, August 12, 1937. (Yuri Kaminsky Collection via Mike Hewitt)

and Dickson Island bordering the Kara Sea. By 1:20 pm Alaska Time the aircraft was still proceeding on course, approximately five hundred miles north of Archangel, having sent a report of a solid cloud layer below their flight level at 7,500 feet. An hour and a half later a transmission from the plane stated a heavier cloud cover was forcing them to divert east around the weather system before attempting to resume their original course.

At 5:20 pm, twelve hours after takeoff, the plane's crew radioed they were back on course, approaching Rudolf Island in Franz Josef Land.[6] All was reported well by the crew. Stations in the United States, Canada and Norway were monitoring the radio signals by then as well.

A little more than seven hours later, at 10:40 am, Levanevsky and his crew reported crossing the North Pole and continuing on the 148 degree meridian toward Fairbanks. They also reported a continuous cloud layer below their cruising altitude of 19,700 feet and the accumulation of small amounts of frost on the airframe. Headwinds of sixty mph were also being encountered at the time.

---

6 Franz Josef Land is a series of Russian islands located at 81 degrees north latitude and the last known landmass between Russia and the North Pole.

As the flight continued southward its radio messages were clearly monitored by the Signal Corps office in Fairbanks. Even with strong headwinds the plane proceeded as planned until it was approximately 250 miles south of the Pole, when the situation began deteriorating. A signal from the plane stated a mechanical problem had developed with the outboard right engine, forcing it to be shut down. Unable to remain at an altitude above the clouds with the resulting loss of power, the aircraft was forced to descend to only 15,000 feet. Levanevsky and his crew were now flying blind in the clouds through icing conditions on only basic flight instruments.

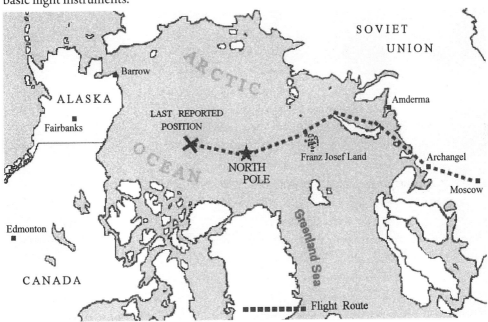

Levanevsky's flight route from Moscow across the North Pole.

With the loss of altitude and increasing amounts of ice accumulating on the airframe, the radio reports from the flight became less and less clear. One garbled message was thought to state they were attempting a landing, but the transmission faded almost immediately. A short time later, a station at Yakutsk in eastern Siberia monitored part of a broken transmission from the aircraft, but heard "all is well." At 5:00 am another garbled message was picked up by the Soviet station at Cape Schmidt. The operator wasn't positive, but thought it was from the plane, asking "how do you hear." As the anxiety increased, radio operators along the route remained glued to their receivers for any hint of communication from the crew. Nothing further was heard.

Even though the plane carried enough fuel to remain airborne for several more hours, with no subsequent contact it soon became apparent the aircraft was probably down somewhere on the ice between the North Pole and the Alaska coast. Some Russian pilots also thought it likely that even if the plane had continued after losing an engine, accurate navigation while flying on instruments would have been extremely difficult, if not impossible. With even a minor navigational error, the crew could easily have become hopelessly lost over the frozen ice cap, eventually being forced to land while low on fuel over rough ice or open leads of water.

The Soviet government and experienced Arctic bush pilots from Alaska and Canada wasted little time in organizing a search for the missing aircraft after it failed to arrive in Fairbanks. Three Pacific Alaska Airways planes were initially hired by the Soviets in Fairbanks and departed early on 14 August. Joe Crosson, a well respected Alaskan aviator, flew a Lockheed Electra over the Brooks Range and along the Canning River to Barter Island on Alaska's Arctic coast, then east for another fifty miles before being forced back by heavy fog. A second plane, a Fairchild 71 on wheels piloted by Murray Stuart, headed northeast along the Yukon and Porcupine Rivers of eastern Alaska to the Arctic coast, then further east for almost an hour before also returning. The third aircraft, a Fairchild 71 on floats flown by S.E. Robbins, flew north to the Brooks Range along Levanevsky's proposed route over the interior of Alaska. After covering more than 2,200 miles in the air, all three aircraft returned without finding any sign of the missing plane.

On 15 August another aircraft was chartered by the Soviets out of Edmonton, Canada. Piloted by Bob Randall, it reached Aklavik on the Arctic coast of Canada and continued west into Alaska, stopping at coastal settlements to question inhabitants about possible sightings of the missing plane.

The same day the Soviet government dispatched the icebreaker *Krassin* to Cape Schmidt on the north Siberian coast, with orders to pickup and deliver three planes to Barrow, Alaska to assist with the search. From there the *Krassin* would proceed north into the polar ice cap and serve as a support base for the search aircraft. Six aircraft in Moscow and three at Rudolf Island in Franz Josef Land were also being prepared to join the search, and an additional aircraft from northern Siberia was dispatched to join the *Krassin* once it reached Alaska. Later that evening, another radio signal on Levanevsky's frequency was monitored by a station at Irkutsk in southern Siberia, but the message was incoherent. The Soviet science station near the North Pole also reported receiving the same weak signal.

Jimmy Mattern, the famous American pilot rescued by Levanevsky years earlier in Siberia, was quickly asked to join the search by the Soviets. More than willing to assist in the search for the fellow aviator who had flown him across the Bering Strait after his crash in 1933, Mattern quickly agreed. He arrived in Fairbanks on 16 August in a twin-engine Lockheed 12A Electra, but remained grounded awaiting a larger supply plane. The bigger Ford 4AT tri-motor was needed for transporting large fuel supplies to a base on the Arctic coast, where the extensive search would be coordinated.

At the same time, Sir Hubert Wilkins, a famous Australian explorer who became well known during several flying expeditions in both the southern and northern Polar Regions, was asked to join the search by the Soviet Ambassador in Washington. He agreed and immediately began preparations for a search from northwest Canada.

By 17 August, Bob Randall flying a chartered Fairchild 82 from Edmonton reached Barrow with promising news. During his flight along the Arctic coast, Eskimos he questioned about Levanevsky's plane at Barter Island, sixty-five miles from the Canadian border, claimed they heard engine noises on the morning of the 13th. They never saw what made the noise and first assumed it was

Bob Randall in front of his Fairchild 82 at Aklavik after returning from the Levanevsky search. (Ohio State University-Wilkins Collection, 33-16)

coming from a boat. After hearing Randall's plane a few days later, the Eskimos thought the sounds could have been from an aircraft. Randall was obviously optimistic after hearing the story and remained in Alaska for another ten days to continue searching.

Wilkins and fellow pilot Hollick-Kenyon departed New York for northern Canada on 19 August in a Consolidated PBY-1 Catalina seaplane purchased by the Soviet government, and the same day a Soviet military twin-engine floatplane arrived at Barrow. The Wilkins expedition planned to initially base out of Coppermine, a central supply point on the northern coast of the Northwest Territories, before eventually moving further west to the village of Aklavik on the mouth of the McKenzie River. From each location they could search north into the Beaufort Sea and east along the coastal areas of Alaska and Canada, utilizing the 4,000 mile cruising range of the twin-engine PBY Catalina.

During the fall season in the upper Arctic, which only lasted through the middle of September, coastal areas were only suitable for water-based aircraft operations, due to a lack of airfields and solid terrain that allowed wheel-equipped operations. Once winter temperatures arrived and froze the waterways and surrounding tundra, flight operations could be continued with ski-equipped aircraft.

By 20 August renewed search efforts were finalized at several of the search locations. Jimmy Mattern had tired of waiting on the refueling plane to arrive at Fairbanks and began searching on his own in the Lockheed Electra. Flying out from the Arctic coast before landing at Barrow, Mattern flew along the 148 degree meridian for four hundred miles. During the long polar flight he encountered fog along most of the route and noted much of the sea ice was unsuitable for a safe landing.

Also on the 20th, the Russian floatplane in Barrow piloted by Zhukov and Randall's Fairchild were undergoing routine repairs and servicing in anticipation of better weather conditions. By then the icebreaker *Krassin* was reportedly nearing Barrow, while Sir Hubert Wilkins was almost at the north coast of Canada on his flight from New York. A Russian radio operator and engineer were also sent north to assist Wilkins with radio communications.

Bad news also arrived on the 20th. The Ford tri-motor sent to supply Mattern's expedition finally arrived over Fairbanks, but a thick overcast at the airfield forced the pilot to land in an open area south of town. Unfamiliar with the spongy muskeg conditions which frequented Interior Alaska, the pilot falsely assumed the terrain was suitable for a safe landing. The soft, wet surface snared the plane as soon as the wheels touched down, flipping it over on its back and causing significant damage. The aircraft was of no further use for the duration of the search.

Soon after arriving at their base in Coppermine, the Wilkins team made their first aerial search on 22 August. Heading northwest they flew out over the ice pack from Prince Patrick Island, reaching a point two hundred miles offshore. The plane's radio was kept on Levanevsky's frequency and monitored continuously for possible transmissions, while the pilot's sent out search messages every half hour in hope of a response. Nothing was heard or sighted of any significance during the thirteen and a half hour flight, with much of the distance flown in limited visibility. During the same time the three aircraft at Barrow piloted by Mattern, Randall and Zhukov, remained grounded because of poor weather conditions, while fourteen Soviet aircraft began supporting the search from the other side of the Pole in northern Russia.

A repeat flight by Wilkins was made the following day on the 23rd, extending even farther over the ice within five hundred miles of the Pole. The flight lasted fifteen hours, much of it in marginal weather conditions requiring extensive instrument flight.

In Sir Hubert Wilkins' account of the search published in National Geographic Magazine in 1938, he described a need for accurate weather reports when operating in the polar region as essential, based on the weather patterns he encountered during his first two search attempts. His subsequent requests for shared weather information from the various governments involved resulted in accurate reports being made available from then on.

On 24 August Mattern was able to depart from Barrow for the first time. Flying in marginal weather he searched north for nearly four hundred miles. Plagued by thick ice fog for most of the flight, he operated mainly on instruments and was limited to monitoring the radio for possible messages from Levanevsky. After concluding the search Mattern continued south to Fairbanks and announced he was halting any further search attempts because of the extreme risks involved. He stated he was particularly concerned about the safety of wheel-equipped aircraft in areas unsuitable for wheel landings, especially when the flights involved extensive instrument flight in hazardous weather conditions. He stated his limited attempts were only able to be completed because of recently installed de-icing equipment. Without it, Mattern was certain he would never have returned safely. He also believed Barrow was not the best choice as a search base, mainly because of the persistent poor weather conditions frequenting the northern coast.

It soon became obvious to the pilots involved that vast areas of the Arctic would never be thoroughly searched with only a few aircraft. The few planes available could only get airborne on a sporadic basis due to weather delays and maintenance requirements, keeping most of the planes grounded in Alaska and Canada. The situation was rumored to be even worse for the Soviet aircraft on the opposite side of the Pole. Search crews still believed that if Levanevsky landed safely in the vicinity of his last transmission, three hundred miles south of the Pole, there was a good chance he and his crew would eventually be found. But if he decided to continue flying past that point, or was not where he thought he was at the time, the situation became much more difficult. The vast area of the Arctic Ocean between Alaska's north coast and the North Pole covered more than 800,000 square miles alone. If Levanevsky had ventured overland into Alaska and Canada, the size of the area in which he might have gone down easily tripled in size.

Even with those seemingly insurmountable odds the search continued with strong expectations of finding Levanevsky and his crew alive. Randall and Zhukov were finally able to takeoff from Barrow on 25 August. The Soviet floatplane covered a short distance north over the ice and Randall flew east toward Barter

Island and Aklavik on his return to Canada. Three large Soviet transports with experienced pilots were also reported to have arrived at Archangel en route to Rudolf Island, to establish search operations there. The following day the *Krassin* arrived at Barrow with supplies for the Russian plane before continuing north. It carried four ski-equipped aircraft aboard, intended for use once the ship was halted by the ice pack.

Weather and maintenance kept Wilkins at Coppermine until the 28th of August, when conditions allowed another long flight northwest over the ice pack. The PBY

Reference map of the Arctic Ocean and vast coastal regions of Siberia and North America.

and its five man crew flew within 300 miles of the Pole before turning back due to a thick fog cover. A new refueling base was established at Aklavik on their return. For the remainder of August weather conditions continued to hamper search efforts at Barrow and Aklavik. On the Soviet side of the Pole, the planes assigned to the search at Rudolf Island were also grounded for extensive periods.

On 1 September Zhukov attempted another search from Barrow, but only flew a short distance north before being forced down on the ice by thick fog. He eventually continued after a few hours and was able to land near the icebreaker

Krassin, where he remained. Wilkins and his crew completed a fourth flight the next day, flying west along the Arctic coast to Barter Island. Soon after arriving a winter storm kept the aircraft grounded for the next five days.

While Wilkins was at Barter Island, possible signal flares from the missing Russian airmen were sighted by a Canadian steamship near Somerset Island, 1,200 miles to the east. The flares were observed in a region rarely traveled, even in the Arctic, instilling a renewed hope the missing Russians were indeed alive and well. After the steamship altered course and investigated the sighting, the flares turned out to be from a local Hudson Bay trader whose vessel had become trapped in the ice. He was rescued soon after.

The PBY Catalina at Aklavik, Northwest Territories, purchased by the Soviet government and flown by Wilkins during the Levanevsky search. Note the Soviet markings painted on the fuselage. (Ohio State University-Wikins Collection, 33-16-4)

During three days in early September Zhukov was able to search an area north from the *Krassin*, approximately 650 miles from the coast. By 7 September the ship was unable to proceed further in the thick ice and its crew began clearing a runway for the four ski-equipped planes. Wilkins departed Barter Island on the same day with improved weather, this time searching farther north than any previous flights. Using a crisscrossing grid pattern the crew was able to cover a much larger area than before. The flight lasted almost twenty-one hours, primarily in the vicinity of Levanevsky's last reported position, but more than half the distance was still flown in the clouds with reduced visibility.

On 10 September the *Krassin* abandoned attempts at establishing a base on the ice and began steaming back south. The ship also reported Zhukov's floatplane had crashed during a water landing and sank, but he was rescued without injury. As a replacement for Zhukov's floatplane the Soviet government announced a twin-engine amphibian would be sent from Siberia as soon as possible to continue the search at Barrow.

Another potential sighting of Levanevsky's missing plane was received on 14 September, when Eskimos at a village inland from Barter Island also claimed to have heard aircraft noises on 13 August. Joe Crosson flew north from Fairbanks in a Lockheed Electra 10C the next day and searched extensively in the area, but found nothing significant. A report of signal flares being sighted off the coast of Barrow by local Eskimos was also investigated, but a subsequent search by Soviet aviator Grazyansky in the recently arrived Sikorsky S-43A amphibian found nothing. Additional flares were sighted by the *Krassin* on 21 September, more than two hundred miles west of Barrow. That area was also searched by the amphibian without success.

Weather conditions continued worsening in the Arctic with the onset of winter and increasing surface ice was making further flights by seaplanes more and more difficult. Even so, Wilkins and his crew managed another flight in the PBY-1 Catalina on 17 September from Barter Island. They reached their furthest point north before being forced to return in heavy cloud cover and icing conditions to Aklavik. It was to be their last search of the year. The winter freeze was quickly closing off landing areas along the coast, making further flights impossible. They departed south for New York a few days later with plans to return once the conditions permitted the use of ski-equipped aircraft.

The four Soviet planes at Rudolf Island continued searching through September, but they had limited range compared to Wilkins' PBY Catalina. Russia's lone plane at Barrow also flew during a few days allowing adequate visibility. With its limited range it never reached a point more than a few hundred miles offshore.

In early October the Soviet government announced it was sending additional aircraft to Rudolf Island in Franz Joseph Land, including four long range multi-engine transports. The amphibian at Barrow completed one additional short range flight before the winter freeze halted further Arctic operations, forcing it to fly south to Fairbanks before returning to the Soviet Union.

A long range search was accomplished on 7 October by a Soviet aircraft from Rudolf Island, reportedly flying past the Pole and south along the 122 degree meridian, using aerial flares during the hours of darkness for reference. Heavy clouds and fog eventually forced the plane to return back to base.

By late November Sir Hubert Wilkins returned north in a twin-engine Lockheed Electra 10, recently purchased by the Soviet government for the search.

Although smaller than the PBY he used previously, the Lockheed Electra was well suited for long range flights. Having an excellent range of 4,000 miles, it had also been fitted with skis for winter operations. Hollick-Kenyon joined Wilkins again as the primary pilot. Their base of operations was initially established at Aklavik, before being moved to Barrow, Alaska on 6 December. With daylight completely disappearing in November as the sun remained below the horizon,

The Lockheed Electra flown by Wilkins and Hollick-Kenyon, at Edmonton, Alberta, on the way north for the Levanevsky search. (NWT Archives, N-1979-003-0825)

it was their intention to only fly during full phases of the moon when maximum ambient illumination would provide enough visibility. The sun would not reappear until late January.

Side view of the Lockheed Electra, showing the new Soviet markings on the fuselage. (NWT Archives, N-1979-003-0827)

By mid-December, four months after Levanevsky and his crew had disappeared, many in the aviation community were convinced the missing Russian airmen were probably dead from lack of food or injuries. Their food supply was long exhausted and no confirmed communication had been established with the plane after it disappeared. Others disagreed and were still optimistic, including Hubert Wilkins. Both Wilkins and Vilhjalmur Stefansson, a world renowned Polar explorer, were convinced the missing men could survive through the winter by hunting sea mammals congregating near open leads in the ice. Eskimos native to the Arctic region had been supplementing their food supply in much the same manner for thousands of years.

Heavy snowfall, low clouds and poor visibility, unfortunately kept Wilkins grounded at Barrow until January. By then deep snowfall and high drifts which accumulated on the surface were a serious hazard to aircraft operations. Wilkins reluctantly returned to Aklavik on 12 January with plans to begin searching anew from there.

The entire operation had by then become almost an effort in futility. Weather conditions were not much better along the Canadian coast, but one flight was accomplished on 16 January during a period of relatively clear weather. The flight lasted nineteen hours, failing to reach a point from which they hoped to find some sign of Levanevsky's plane. After their return, a mechanical problem on one of the engines kept the Lockheed Electra on the ground until a replacement was found and installed in late February.

The next flight wasn't feasible until March 3rd and Wilkins decided to test the new engine by searching inland into the interior of Alaska. Wilkins and Hollick-Kenyon first searched west along the north side of the Brooks Range through half the width of Alaska, before turning back east to Herschel Island on the Canadian coast. After reaching Herschel Island they reversed course again, flying west to Alaska's Colville River, then back east once more along the Arctic coast to Aklavik, covering more than 2,600 miles during the flight.

Clear skies again prevailed over the interior of Alaska the next day, allowing another search attempt on the south side of the Brooks Range that covered 1,800 miles. After searching extensively during the two days of clear weather without spotting any evidence of the missing plane, Wilkins was convinced Levanevsky did not reach the Alaskan mainland.

Wilkins set out again on 10 March after waiting several more days for improved weather, this time searching further northeast over the Polar ice between Prince Patrick and Ellesmere Islands. As had often occurred on previous attempts, the visibility was diminished for most of the flight by low lying haze and a solid cloud deck that forced them to return before intended.

The next and final flight by Wilkins and Hollick-Kenyon was flown on 14 March in favorable weather conditions. They searched further north than ever before, reaching a point less than two hundred miles from the North Pole. More than 3,000 air miles were covered in almost twenty hours of flying, with excellent visibility throughout the trip for the first time. Before they could fly again, two additional flights planned during the month were suddenly cancelled by the Soviet government.

The abrupt notice was received on the 15th in the form of a telegram from the Soviet Ambassador in Washington, informing Wilkins the search for Levanevsky was being terminated. The message stated no future attempts to locate Levanevsky and his crew were to be launched from the American and Canadian side of the Arctic. Since the search by Wilkins was being funded by the Soviet government, he had little choice in the matter and returned to New York on 25 March. His search for Levanevsky had ended.

The Soviet government announced that the search for their missing airmen would continue with Soviet aircraft from Rudolf Island, utilizing good flying weather during the upcoming summer months. They explained that few attempts were made by their aircraft during the winter because of poor weather, although it was later revealed six of their aircraft were lost in accidents. Their search would concentrate in an area between the Pole and the Greenland Sea, where the Arctic currents would have carried the polar ice following the probable ditching of Levanevsky's plane. Further details concerning the subsequent search for Levanevsky were not revealed, nor were reports of any significant findings.

During the seven month period from August 14, 1937 to March 14, 1938, more than 200,000 air miles were flown over the Arctic by search aircraft operating from Alaska and Canada. Sir Hubert Wilkins estimated his flights alone covered 170,000 square miles.[7] Thousands of additional miles were flown by support aircraft carrying supplies to Aklavik, Coppermine, Barter Island and Barrow. Twenty Soviet aircraft were reportedly involved in the search from the opposite side of the Pole.

Following the disappearance of Sigismund Levanevsky and his crew, the Soviet government's internal investigation, which remained secret for many decades, revealed critical mistakes in maintenance and crew training. It was determined much of the maintenance performed on the aircraft was conducted without proper supervision and the crew had not been permitted to participate in any of the technical modifications. On several occasions problems with engine exhausts and radiators occurred, posing a danger of in-flight fire that was not adequately resolved. Inadequate landing gear for the increased aircraft weight was not noticed or corrected until shortly before departure. Crew flight training should have been at least thirty hours, but only ten and a half hours were flown. Those hours in the air did not include long distance instrument flight or operations with supplemental oxygen, and ground training was insufficient for the new aircraft systems.

---

7 Sir Hubert Wilkins estimate of 170,000 square miles was based on good visibility for approximately half their total flight time, allowing five miles of observation on each side of the aircraft.

Finger pointing ensued in the Soviet government, resulting in several scapegoats being identified, including a radio operator in Siberia who was given a twenty year prison sentence for failing to adequately monitor radio transmissions. Other individuals disappeared entirely or died in mysterious accidents.

On an international level the mystery of Levanevsky and his missing plane faded from public interest, until an eyewitness account in 1938 seemed to substantiate earlier reports of a sighting near Barter Island. In the spring of 1938 an Eskimo arrived at Barrow describing a strange sighting he had witnessed the previous fall while camped on Oliktok Point, midway between Barrow and Barter Island on the Arctic coast.

According to the Eskimo's account of the incident, he and seven other members of his group heard what sounded like an outboard motor offshore from their camp. He said the object was not visible at first and was believed to be a boat, but after searching through binoculars a shape was seen moving above the horizon at a fast rate of speed. The Eskimos observed the object moving east between Thetis and Spy Islands, four miles away, then suddenly disappear into the water with a large splash. No further sounds were heard. After investigating the area in their boats, they located an oil slick on the surface of the water.

The Eskimo told his story to a Signal Corps operator and a local missionary living in Barrow, who both considered the account believable. The report drew little interest outside the community until an expedition arrived at Barrow in August 1938 to erect a memorial in honor of celebrities Wiley Post and Will Rogers, who died in the crash of a Lockheed Orion three years earlier. As leader of the memorial expedition, Homer Flint Kellems was first skeptical of the Eskimo's story, but after further inquiries became convinced of its authenticity. The Eskimo, who first reported the strange sighting in the spring of 1938, had shown an entry in his diary to the Signal Corps operator and missionary at the time, describing the incident. Kellems' investigation also confirmed the sighting with other members of the hunting party present at Oliktok Point. The owner of a Trading Post at Beech Point, seventeen miles east of Oliktok, verified the same individuals reported the sighting to him only a few days after it occurred in August 1937.

Once Kellems became convinced the Eskimos were telling the truth, he organized a search of the area where the object disappeared. Operating from a small power boat, Kellems' team dragged weighted grappling hooks across a grid pattern covering a thirty-six square mile area between Thetis and Spy Islands. Using a simple compass to maintain a fixed course through the shallow water was difficult and at times almost impossible, but they made a determined effort. Engine problems and strong currents added to the difficulty, leaving gaps in the search grid, but a large magnetic disturbance was recorded, possibly triggered by one or more large metal objects, such as the engines from the missing plane. Unfortunately, the disturbance occurred while the boat was adrift and they were unable to relocate the exact position before leaving with the onset of winter. Even so, each team member was certain they had been near the lost plane.[8]

It is understandable why the Soviet government took little interest in organizing another search for Levanevsky after the summer of 1938. By then a civil war in Spain was receiving the Soviet government's full support and funding, including a good deal of their aviation assets. Open conflict with Japan and Finland in 1939 and early 1940, followed by their entry into World War II with Nazi Germany in 1941, plunged the Soviet Union into years of devastating conflict. Almost immediately after World War II ended, a Cold War between the Soviet block and the western powers began, lasting nearly five decades. Levanevsky, the "Soviet Lindbergh," was all but forgotten as his disappearance was pushed into the back pages of history.

Whether Levanevsky and his crew crashed somewhere on the polar ice cap, in the ocean or in a remote stretch of wilderness, survived for any length of time or perished upon impact, might never be determined. The few aircraft involved in the search could only have adequately covered a small fraction of the area in which Levanevsky vanished. Even with binoculars an individual's ability to see objects on the surface from the air is limited. A large four-engine transport like Levanevsky's, if intact and not covered by snow or partially buried, would still appear as a tiny speck at a distance. The likelihood of the plane or crew being found in such a vast expanse covering hundreds of thousands of square miles, could only have been accomplished by the sheer luck of a passing search aircraft.

The northeastern portion of Alaska from the Brooks Range to the Arctic coast, which alone covers more than 30,000 square miles, was only briefly searched from the air. History has shown that in many cases aircraft which disappeared in much smaller areas have never been found. Other areas of the Arctic cover much greater territory. The rugged, almost endless stretches of mountains, glaciers, lakes, tundra and forests of Alaska, Canada and Siberia still hold many aircraft mysteries today.

Interest in the mystery of Levanevsky surfaced again in 1987 when an American citizen was researching a missing U.S. Army Air Forces B-25 Mitchell bomber interned in the Soviet Union during World War II. During Walter Kurilchyk's research, he was given a copy of the 1938 report sent to the Soviet Embassy by Dr. Homer Kel-

---

8 Dr. Homer Flint Kellems's search is documented in "We tried to Solve an Arctic Mystery," published by *The Alaska Sportsman* in June 1940, and in a report he sent to the Soviet Embassy in 1938.

lems on the search for Levanevsky's plane. The letter had only been recently located in Soviet archives and poor relations between the Soviet Union and the United States prompted an official of the Soviet government to ask Kurilchyk to investigate the validity of the report through unofficial channels. The Soviet official requesting Kurilchyk's assistance was General Baidukov, Colonel General of the Soviet Air Force and co-pilot during his country's first non-stop trans-polar flight in June 1937. He was also an old friend of Levanevsky. Walter Kurilchyk would later publish a book on his findings after organizing an expedition of his own.[9]

With the assistance of Arco Alaska, Inc., a large conglomerate with oil interests on the north slope of Alaska, Kurilchyk was able to conduct an aerial search of the Thetis and Spy Island area in March 1987. That search resulted in a second attempt with boats and divers in August 1987, using magnetometers for detecting anomalies on the ocean floor. Several strong readings were recorded on a line between Thetis and Spy Islands. An aerial magnetic survey of the area conducted in the late 1940s was also uncovered in old files by Kurilchyk that revealed several magnetic anomalies near Spy Island.

Persistent problems with equipment and ocean currents, similar to those experienced by Dr. Kellems in 1938, haunted Kurilchyk's expedition. Eventually the search was abandoned without recovering evidence of the missing plane, but those involved were confident they were looking in the right area.

A second expedition to Oliktok was conducted in April 1990, this time by dragging sleds with magnetometers over the frozen surface. Similar anomalies to those found in 1938 and 1987 were observed, but no conclusive evidence was found establishing the presence of aircraft wreckage or other manmade objects. No further expeditions were initiated.

In September 1987 other clues to Levanevsky's disappearance were being investigated several hundred miles east of Oliktok, near the village of Old Crow in Canada. It was there the Arctic Institute of North America conducted a survey of an old camp site from the 1930s, first discovered in 1967 by occupants of a small aircraft which crashed nearby. The strange site contained a small wooden enclosure containing skeletal remains, old style metal fuel cans, a woodpile and old food tins, possibly from Levanevsky.

A scientific investigation team found indications of additional metal objects under the ground with the use of metal detectors, but the Canadian government would not allow an excavation until an archeological survey could be conducted. Tree samples taken by the team from previously cut timber around the site suggested the date of the camp somewhere between 1935 and 1939.

---

9 *Chasing Ghosts*, published in 1997 by Aviation History Publishing

An archeological team returned in July 1989, confirming evidence that several individuals had spent a winter at the site. Excavations conducted around the camp revealed old tin can fragments, broken glass, different sized notched and cut logs and a spent rifle case. Samples were also taken of the skeletal remains, tree cores and collapsed wooden enclosures for analysis.

As promising as the site was initially believed to be in solving the Levanevsky disappearance, it eventually proved to be a false lead. Skeletal remains from the camp were determined to only be from caribou carcasses and the wooden enclosures, first thought to be an old shelter and hastily made coffin, were in fact remnants of an old food cache and wolverine trap. Tree core samples revealed it had been occupied in the late 1920s or early 1930s, much earlier than Levanevsky's disappearance. The site was later confirmed by a local elder as once used by prospectors as a food and supply cache in 1929 and 1930.

In March 1999, a pre-drilling sonar survey of Camden Bay by Arco Alaska, Inc. revealed a large anomaly on the sea floor, approximately sixty feet long in the shape of a large aircraft fuselage. Camden Bay lies a hundred and twenty miles east of Oliktok Point and only thirty miles west of Barter Island. The sonar survey prompted an investigation into whether the image might be the long lost Russian transport.

A second survey was conducted in the thirty foot deep water, using side-scanning radar and a video camera to try and identify the object. Rough seas and extremely murky water did not allow the team to find anything conclusive. The same scientific team returned again in May 2000, this time using a remote controlled camera lowered through the ice. Even though the water was much clearer than the first attempt, the camera failed to locate the image or find anything from an aircraft. After reanalyzing the sonar scan and verifying the correct coordinates, the team determined their search had been off by almost a hundred yards, leading to another search in the spring of 2001. This time the sixty foot image was located, but it turned out to be only a large rock formation buried in the silt.

Possible aircraft wreckage from Levanevsky's plane is also believed by some researchers to be located along the shore of Ellesmere Island, off the northwest coast of Greenland. Several aerial searches have recently been organized in that area under the assumption Levanevsky would have flown toward the nearest land mass after his aircraft developed engine problems. Ellesmere Island is five hundred miles nearer the point of Levanevsky's last radio message than the north coast of Alaska.

Rumors have persisted over the fate of Levanevsky and his crew since they first vanished in 1937. Some theorize the plane was sabotaged by Stalin because of jealousy over Levanevsky's fame or disgust with his aviation failures and resulting embarrassment to the Soviet Union. Others claim the flight was a hoax perpetuated by Stalin after first having Levanevsky killed, when it was uncovered he intended to defect after arriving in the United States.

Only the passage of time can reveal the truth to Levanevsky and his crew's mysterious disappearance. For whatever remains of the missing plane and the six lost airmen is still out there, somewhere in the Arctic.

Chapter Two

February 5, 1943

Secret Cargo

A solid overcast hid the heavily forested hills from the lone military transport flying overhead, leaving only the tops of the high, snow-capped mountains poking through the thick clouds into the night sky. Flying the airway system of the Northwest Staging Route between Fort Nelson and Fort St. John, British Columbia, a U.S. Army Air Forces C-49K was paralleling the rugged Rocky Mountains of western Canada stretching in a jagged line off the right wing. Foothills and lesser mountains, invisible through the thick overcast, tapered away for hundreds of miles off the opposite wing into a large expanse of open tundra extending into the distance. Only a few communities dotted the remote wilderness below, mostly small towns and outposts scattered along the well used, narrow dirt road running through Canada, linking Alaska and the continental United States.

The Douglas C-49K was assigned to Headquarters and Headquarters Squadron of the Air Transport Command's Alaska Wing at Edmonton, Alberta, and as such performed administrative missions from various airfields along the Northwest Staging Route. On this day the plane was carrying three crewmembers, eight passengers, an unidentified cargo and sacks of mail on a service mission from Whitehorse, Yukon Territory to Edmonton, with four scheduled stops en route at Watson Lake, Fort Nelson, Fort St. John and Grand Prairie. A strong weather system inundating the region from central British Columbia to Interior Alaska with extreme cold temperatures and heavy snow showers had grounded all flights for the previous two days, until a gradual improvement finally developed on 5 February.

After originally departing from Whitehorse in the early afternoon on an instrument flight clearance, the military C-49K arrived at Watson Lake, Yukon

Territory ninety-two minutes later, exchanging some passengers and mail before continuing southeast to Fort Nelson, British Columbia. Arriving without incident, the crew spent slightly more than four hours on the ground transferring mail, servicing the aircraft and taking time to enjoy a warm meal. The passengers relaxed or enjoyed a meal as well, except for two of the four soldiers who remained in shifts with the cargo.

Airfields and navaids along the Northwest Staging Route.

One of many C-47 cargo planes that passed through Fort Nelson on the Northwest Staging Route, April 1943. (Burt Prothoro via Blake Smith)

The flight lifted off the runway from Fort Nelson later that evening on an instrument flight plan, climbing through the clouds to an altitude of 8,000 feet. A time en route of one hour and twenty-five minutes was estimated to the next destination of Fort St. John, British Columbia, one hundred and sixty-six nautical miles further southeast. Eight minutes after departure the pilot reported over the Fort Nelson Radio Range and turning onto the Amber 2 airway. This placed the aircraft on course a few miles southwest of the airfield where it would continue along the airway system.

The next position report was not due until passing over Beaton River, approximately mid distance along the airway. As far as anyone on the ground was concerned the flight was proceeding normally. No distress calls were received, no position reports, no garbled transmissions, no routine or unusual communications of any kind. The aircraft simply did not arrive as scheduled.

C-47 taxiing at Fort Nelson, May 1943. (Burt Prothoro via Blake Smith)

Two and a half hours after the C-49K left Fort Nelson and an hour after its estimated time of arrival at Fort St. John, it was officially reported as missing by the military. A communications alert was immediately issued to all stations along the route in the chance the aircraft had diverted to a different location. When no new information could be established regarding its whereabouts, an area wide search was immediately organized. It seemed as if the aircraft had vanished without a single clue.

The missing plane was a relatively new model Douglas DC-3, given a C-49K designation by the U.S. Army Air Forces after being acquired from Trans World Airlines at the outset of World War Two. It was one of many Douglas DC-3 variants requisitioned by the military from civilian sources to fill the critical demand for transport and cargo aircraft at the time. Other civilian DC-3s acquired during the war were given military designations from C-32 to C-117. A total of 138

C-49s alone were used by the U.S. Army Air Forces during World War Two, all configured with 1,200 hp Wright Cyclone R-1820 radial engines. The aircraft was capable of a maximum speed of 212 mph, a cruising speed of 192 mph, a range of 2,000 miles and a ceiling of 20,800 feet.

The military did order thousands of its own version of the DC-3 as well. Eventually manufactured in great numbers, the aircraft was officially known as the C-47 Skytrain. Affectionate names by other countries and their various branches of service also referred to the flying workhorse as the Gooney Bird, Skytrooper and Dakota. Although C-47s were almost identical in appearance to DC-3s, the many subsequent variations included several structural, electrical and engine modifications to meet more stringent military mission requirements.

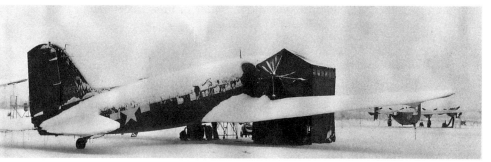

An Air Transport Command C-47 being preheated for flight at Ladd Field near Fairbanks, Alaska. (Randy Acord)

A pilot, co-pilot and a flight engineer were the only crew aboard the missing C-49K. Lieutenant Colonel Mensing was the designated pilot-in-command and a highly experienced former airline captain who had flown with Northwest Airlines before the war. The co-pilot was Lieutenant Charles Atwood and the flight engineer Sergeant Sam Schilsky. Four of the eight passengers being carried with cargo and mail were military personnel and the other four were civilian technicians under contract with the U.S. Army.

Weather conditions at the time of the plane's disappearance were considered favorable for instrument flight conditions. A high overcast cloud layer at 5,000 feet and a predominate visibility of eight miles was forecast along the route. Intermittent snow showers and isolated areas of lower ceilings were also predicted with occasional light turbulence, but nothing of significance that would adversely affect instrument flight conditions. Winds at their flight level of 8,000 feet were expected out of the southeast at thirty knots.

Instrument navigation between Fort Nelson and Fort St. John was accomplished by tracking a low frequency airway system being beamed from three radio range stations dispersed along the route. Pilots flying between the airfields would initially navigate outbound from the Fort Nelson Radio Range, following the Amber 2 airway southeast past the Beaton River Radio Range until intercepting the northwest leg of the radio range station at Fort St. John. Once the pilots verified they were over the radio range station at Fort St. John, an instrument approach procedure would be initiated to land at the airfield.

The radio ranges at Fort Nelson and Fort St. John were part of the newly established Northwest Staging Route linking the United States with Canada and Alaska during World War II. The system provided an essential air corridor for supplying men and equipment to strategic U.S. military bases in Alaska during the war and was the same route used by thousands of lend-lease aircraft flown to the Soviet Union. Officially beginning in Great Falls, Montana and extending more than 2,000 miles northwest to Fairbanks, Alaska, the Northwest Staging Route roughly followed the same course through Canada as the Alcan Highway, completed in 1942.[1] The air route was interlinked by dozens of communication facilities, major airfields and smaller emergency airfields; all manned by thousands of military and civilian support personnel. When search and rescue operations were required in the region, those same individuals and facilities became essential for organizing air assets and aircrews involved in the search for a missing aircraft.

Following the disappearance of Colonel Mensing's aircraft on the evening of 5 February, all assets of the Air Transport Command's northwest region were notified to prepare every available aircraft for a search at first light the following morning. A second military transport also disappeared the same night at approximately the same time, but farther northwest in the Watson Lake region of the Yukon Territory. The second aircraft, a C-47, was carrying a crew of three and three passengers when it vanished while on an instrument approach into Watson Lake.

By the morning of 6 February numerous aircraft from both American and Canadian forces were waiting on improved weather conditions to initiate a search for both aircraft, but low ceilings and poor visibility in the area forced many of them to remain grounded for extended periods. Five planes at Edmonton and six at Whitehorse could not depart until late morning, and aircraft at Fort Nelson, Watson Lake and Fort St. John were delayed even further. Other planes dispatched to assist with the search were still en route from various locations as far as Fairbanks, Alaska and Great Falls, Montana.

Three main search bases were initially established at Fort St. John, Fort Nelson

---

[1] The Alcan or Alaska-Canada Highway was constructed as a military road system linking the continental United States with Alaska following the Japanese attack on Pearl Harbor. Covering fourteen hundred miles of wilderness from Dawson Creek, British Columbia to Delta Junction, Alaska, the road was completed in a record time of only eight months and twelve days.

and Watson Lake, with the majority of available aircraft based out of those airfields. All the aircraft involved in the search mission carried extra survival equipment that could be air dropped to survivors, as well as additional crew members to serve as spotters. In addition to the extensive aerial search that was organized, special ground teams were also outfitted for search operations into heavily forested areas not easily observed from the air.

Additional communication attempts with both missing planes from radio facilities along the Northwest Staging Route were again conducted on 6 February, without success. All communication outlets also confirmed no radio transmissions were received from Colonel Mensing's aircraft since its last contact over the Fort Nelson Radio Range or from the other aircraft after arriving over Watson Lake. Further attempts by search aircraft throughout the day failed to receive any reply from either missing plane, even though both carried a "Gibson Girl" survival radio for emergency situations.

Weather conditions frequenting the region again interfered with search efforts on 7 February, allowing only a limited search around Fort Nelson. As frustrating as the conditions were to the search teams waiting on improved conditions, the weather unexpectedly changed for the better by the following morning, allowing nineteen aircraft to get airborne for extensive periods. For the first time aircrews were able to cover a wide area around Watson Lake and both sides of the airway for several miles from Fort Nelson to Fort St. John. Nothing was found.

Over the next two months the continuing search missions were plagued with obstacles. Persistent low ceilings, poor visibility, high winds, extreme cold temperatures, vast areas of thick timber and seemingly endless mountain ranges made search efforts along the Northwest Staging Route extremely difficult. The C-47 which disappeared in the Watson Lake area was found on 23 February, only a few miles off the end of the runway in heavy timber, while the search for the missing C-49 continued. The two Northwest Airlines contract pilots on the C-47 were killed when it crashed in the trees on approach to the airfield, but the other four occupants were found alive, although in various states of injury.

The search for Colonel Mensing's aircraft continued for another eight weeks. No public mention was made of the cargo, but there was a strong influence to use all means necessary in finding the plane. Reluctantly, after finding no evidence of the C-49 aircraft or the eleven occupants, the search was officially terminated on 6 April. Hundreds of flight hours involving dozens of aircraft and numerous support personnel were spent searching for the missing plane, covering tens of thousands of square miles by ground and air, to no avail.

In the Aircraft Accident Report issued by the Army Air Forces, the exact nature of the accident could not be determined due to a lack of evidence. Analysis of the information that was available determined the weather conditions were worse than forecast, but adequate for instrument flight. While on the ground at Fort Nelson the co-pilot had received a weather update from the base Operations Officer and both mutually agreed the weather was sufficient for safe instrument operations. Likewise, the crew did not report any mechanical problems or other deficiencies with the aircraft.

The Accident Investigating Officer believed the loss of the aircraft was in part caused by inadequate radio equipment on board the plane. His assumption was based on the likelihood the pilots were unable to establish radio contact for an instrument approach at Fort St. John and continued flying in an attempt to find more suitable weather. That belief was shared by other pilots in the Air Transport Command, many who experienced episodes where they were unable to make radio contact with airway and airfield control facilities because of poor communication equipment and weather interference. This was a recurring problem throughout the war.

Accurate weather reports along the Northwest Staging Route were also found to be a problem, but not because of a lack of trained personnel. At the time of the aircraft's disappearance weather information was only being transmitted between stations in code. Unfortunately, the decoding manuals were not yet available at many of the airfields, making accurate weather forecasting nearly impossible. This problem was soon rectified after the two transport planes disappeared on 5 February.

A possibility also existed that Colonel Mensing's plane experienced a mechanical problem or navigational error, causing it to crash unexpectedly before a distress call could be transmitted, but neither of those theories could be substantiated. Until the missing aircraft could be found and examined, the circumstances of the accident would remain a mystery.

As time passed along the Northwest Staging Route the military in the region again focused its full attention on the war effort, eventually pushing the aircraft's disappearance into the archives of history. Rumors persisted for a while, but like many others lost during times of conflict, the plane and crew were simply written off and forgotten, missing as casualties of war.

Five and a half years after the C-49K disappeared on a routine flight between Fort Nelson and Fort St. John, the remains of a military style transport plane were found on a remote mountainside seventy-six miles southwest of Fort Nelson. The wreckage was discovered during the fall hunting season in September 1948, when a guide from Fort Nelson and his American client from Oregon spotted a bright object on a far hillside they had been glassing for sheep. At first they assumed it was only the sun reflecting off a patch of snow or ice, but it seemed out of place on the rock covered slope and continued to draw their attention. After hiking several miles to the mountain and up a steep moraine to investigate the sighting, they were surprised to find the mangled wreckage of a twin-engine aircraft and skeletal remains of several individuals. Among the broken and burned pieces of metal they also found a torn canvass mail sack and a few scattered letters dated from the 3rd and 4th of February 1943.

Hunting guide Georgie Behn in front of the wreckage he discovered in 1948. (Georgie Behn)

Realizing their find was a previously unknown crash site, the guide and client hurried back to their camp on Tuchodi Lakes to inform the other members of their hunting party. It was decided one of them would leave as soon as possible and notify the proper authorities of the find, which was done the following day. When a detachment of the United States Air Force (USAF) Air Transport Wing assigned at Fort Nelson learned of the find, they immediately forwarded the news to the Northwestern Air Command in Edmonton, Alberta. After consulting with officials from the Royal Canadian Air Force (RCAF), it was decided a recovery would be initiated under the direction of American forces with the assistance of a Canadian unit stationed at Fort Nelson.

A hastily organized team was flown into Tuchodi Lakes the next day on 21 September by a RCAF floatplane. From there a base was established on the lake

Chapter 2—Secret Cargo

shore approximately seven miles south of the crash site. The recovery team consisted of several military personnel from the USAF and RCAF, a Canadian constable from the Provincial Police, the two individuals who found the wreckage and an additional guide from the hunting camp who had remained at the lake.

The team departed for the crash site early the next morning using saddle horses provided by the hunting guides for transportation. Those were augmented by several pack horses for hauling supplies and bringing out any human remains

Oregon hunter who accompanied Georgie Behn, beside a landing gear assembly down slope from the main wreckage. (Georgie Behn)

and cargo they could recover. After several hours of travel the team reached an alpine meadow below the wreckage and set up a base camp before continuing up the mountain a short distance to the point of impact.

After reaching the wreckage the team was surprised by the extent of damage to the aircraft which left mostly small pieces of metal littered along a 350 yard area. It was apparent from the pattern of debris that the aircraft had hit a higher ridge above the wreckage site while flying on a westerly heading at approximately 8,500 feet, well north of the southwest leg of the Fort Nelson Radio Range they should have been over. Upon impact the aircraft exploded into multiple pieces, throwing the demolished airframe backwards down the steep moraine. One broken wing and a twenty-five foot section of the torn fuselage were the two largest pieces remaining. Most of the visible wreckage consisted of small pieces of metal that had been scattered about the slope from the high impact collision and ensuing rock slides. Even the engines had been crushed by the force of impact, making them almost unrecognizable among the other debris. Some pieces had been

Front engine assembly torn from the aircraft during the crash. (Georgie Behn)

View of the main area of wreckage from the C-49K found on the high mountain slope. (Georgie Behn)

partially burned, in all likelihood by the instantaneous ignition of fuel during the crash. Shortly after impact and probably throughout the ensuing years, rocks and dirt had fallen from the higher moraine, scattering and covering many pieces of debris under several feet of earth.

Little remained of the aircraft structure to identify the registration number, but skeletal remains and clothing found below the broken wing and fuselage section provided positive confirmation it was the missing aircraft. Partial remains were found from all eleven individuals on the aircraft, and eight of them were identified on site through clothing and personal items still on the bodies. No cargo, other than mail, was apparently identified.

All the human remains were collected and carefully wrapped before being loaded on the pack horses for transport down the mountain, as well as personal items and several pieces of official mail that had been found. A thorough search of the rocky slope continued over the next two days, but no documentation was forthcoming on what they were still looking for or what they found. Was it missing cargo, or was the team just being thorough in recovering all possible remains?

In any case, once the substantial team of men, saddle horses and pack horses returned to Tuchodi Lakes, the recovered items were flown to the military base at Fort Nelson. From there the human remains located at the crash site where sent to the Mortuary Office at Great Falls, Montana for further identification, before eventually continuing onto their families for burial. Other recovered items were stored in a secure building for further processing. No official paperwork has been uncovered listing the identity or eventual destination of the items.

Guide Georgie Behn returning to Tuchodi Lakes with one of the mail sacks found in the wreckage. (Georgie Behn)

Following the recovery of the remains at the crash site southwest of Fort Nelson, a second Accident Investigation Board convened concerning the missing

C-49's discovery. However, because of the extensive destruction to the aircraft during impact and lack of evidence available at the scene, no cause factor could be determined. Without adequate information the board could not draw a realistic conclusion or provide recommendations.

Why the aircraft crashed so far off its intended flight path still remains a mystery. The point of impact was well west of the radio range instrument approach course into Fort Nelson and further west by another thirty miles than the Amber 2 airway course to Fort St. John. Because the wreckage was found at the same approximate altitude the plane had been cleared to fly, it is likely radio navigation

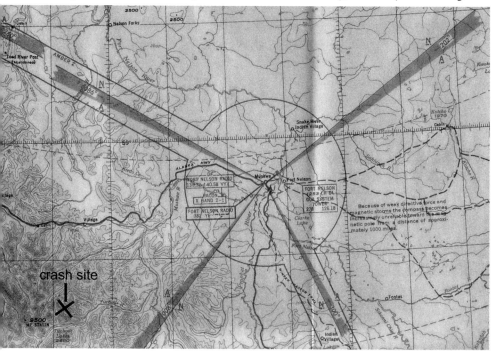

1940s World Aeronautical Chart showing the Fort Nelson Radio Range and airway system between Watson Lake and Fort St. John.

and communication was lost shortly after intercepting the Fort Nelson Radio Range. Without a reliable signal to track the airway beam and a solid overcast that night blanketing the ground, the pilots would have had no way to identify their position. Probably in an attempt to return to Fort Nelson they then crashed into higher terrain.

Testimony from the Investigating Officer who accompanied the recovery team did provide some insight as to why the wrecked aircraft had not been found during the previous five years. In his opinion there was not enough left of the aircraft structure to draw attention from the air and the small pieces of metal still lying above the surface of the moraine could have been easily mistaken for snow or ice, which still existed on some areas of the mountain even in September. Hunters had also been in the same area for years and only the sun reflecting of a piece of exposed metal at the right time in the fall of 1948 sparked an interest in the site.

Other factors also contributed to the wreckage remaining hidden for so many years. The crash occurred in winter when heavy snowfall already covered the mountain slopes and new snow continued blanketing the region for weeks following the accident, effectively hiding any evidence of the wreckage. By summer even the scattered pieces of metal were in large part camouflaged by ensuing rock slides which occurred after the crash, by which time the search for the missing aircraft had already been terminated. During the subsequent years the mountain slope was almost always covered by winter snow and was only exposed for a few months in late summer and early fall. Unless an aircraft flew over the mountain at a low level when the snow was melted, and the occupants were looking down at the site at that exact instant, a visual sighting from the air would have been impossible.

A bigger mystery unfolded almost as soon as the wreckage was found, but whether it was based on wishful thinking or hard evidence has never been verified. The same day the recovery team reached the crash site on 22 September, a story in the Edmonton Journal reported the plane had been carrying more than 200,000 dollars in currency and 400 pounds of gold bullion when it disappeared. It was a fantastic story sparking rumors of secret government treasure that was soon picked up by major newspapers across the continent. Could it be items recovered at the crash scene had been leaked to the media?

The USAF immediately discounted the report, but the Edmonton Journal carried a follow up story the next day, claiming there was nearly 500,000 dollars in gold and currency on the aircraft when it crashed. A special dispatch from the paper said five Oregon hunters had found the wreckage along with thousands of dollars in currency, which was described in detail to the reporter by a member of the hunting party during a stopover in Fort St. John while returning home.

The incredible tale persisted for several more days and it was reported the FBI was even en route to investigate. In the mean time the USAF denied the aircraft carried any cargo at all, but further claims were still being carried by the Edmonton Journal. A wife of one of the hunters, who had talked to her husband by telephone shortly after the discovery, said he claimed there was more than a million dollars in gold and currency still left at the crash site and the recovery team was still in the process of digging it out. So who was telling the truth?

When members of the hunting party were contacted by the newspaper on their return flight through Edmonton, the story was suddenly changed to say the hunters were only under the impression the plane carried a valuable cargo of gold bullion and cash, but had not seen any of the cargo themselves. One of them reportedly claimed it was in fact USAF officers who had told them it was carrying a large amount of gold and currency. A follow on story changed the circumstances yet again, now reporting members of the hunting party had only been told of the secret government cargo by local Canadians at Fort Nelson and not by any USAF personnel. The big game hunter who originally found the wreckage was later quoted by a newspaper article as saying they only found human remains and a small amount of currency in the victims' clothing, and no evidence of any cargo of gold or cash. But why the sudden change in the story?

The truth to the rumor of gold bullion and currency on board the wrecked C-49 has never been substantiated. Perhaps it was only a tall tale instigated by the local population or the exaggeration of an over-zealous reporter. It's possible a rich cargo of gold and currency could have been carried aboard the aircraft, but if it was still with the wreckage in 1948 why would the cargo be covered up by the U.S. government after the news of its discovery first materialized?

When the Accident Investigation Board met three weeks after the missing plane was found, a member of the board did inquire whether any cargo had been located at the crash site, but did not elaborate further on the contents. In spite of any information which might have motivated the board member's inquiry, testimony given by the Investigating Officer who had accompanied the recovery team to the wreckage, stated there was nothing to indicate cargo was aboard the plane. No further inquiries were made by the Investigation Board concerning the cargo, even though newspaper accounts of gold and currency were already well publicized. It seems odd that the rumors, if they were rumors, were dismissed so easily.

No copies of the aircraft's cargo manifest were included in the official Accident Report, even though it was documented that at least some cargo in the form of mail sacks was being carried. A cargo manifest for the missing flight should have been on file and still available to the Investigating Officer in 1943, as well as the second Investigation Board after the aircraft was found in 1948.

Perhaps a valuable cargo was indeed aboard, only to be found and salvaged before the hunters stumbled on the plane in 1948, or was still in the wreckage in 1948 as first reported shortly after the discovery. Whatever the true circumstances, the story of secret gold and currency being carried on a military transport in the winter of 1943 remains a zealous tale.

Over the years stories of lost treasure have persisted in other aviation disasters. The thrilling reports of a secret cargo in the wreckage of Colonel Mensing's aircraft near Fort Nelson were similar to other rumors of a cargo of gold bullion at a remote crash site in Alaska in 1948. It was during March of that year that a chartered Northwest Airlines DC-4 impacted one of the rugged peaks of the Wrangell Mountain Range near Gulkana, killing all thirty occupants onboard instantly in a fiery explosion.[2] What was left of the plane and any cargo fell several thousand feet onto a narrow glacial field and was never recovered.

During the ensuing decades repeated stories of a cargo of gold bullion aboard the Northwest Airlines plane spawned numerous attempts to reach the wreckage, but all were unsuccessful until the late 1990s. In July of that year a pair of aviation enthusiasts who had been researching the crash finally located a small amount of wreckage in a debris field that had been carried away from the mountain by the slow moving glacier. Most of the remaining aircraft debris had become entombed and hidden deep within the ice over the previous five decades, where it will probably remain for years to come.

Even though no evidence of a gold cargo was found at the glacier in Alaska, rumors of gold still persist over the mysterious crash in British Columbia. Among adventurers dreaming of a lost fortune, the rumors will never completely die, for what is a better mystery than a lost fortune waiting to be recovered?

---

2 Details of the Northwest Airlines crash are discussed in a previous book by G.P. Liefer, titled *Broken Wings: Tragedy and Disaster in Alaska Civil Aviation*.

Chapter Three

June 19, 1943

All Reports Normal

A pink sunrise encased the high mountains of the Alaska Range as the bright morning rays of light filtered through the distant peaks. Scattered clouds lay over the military airfield near the city of Fairbanks, fading into the horizon. Almost endless daylight was covering the region during the spring and early summer as the sun approached its zenith over the Northern Hemisphere. Only a short period of pale twilight occurred each evening as the sun moved briefly below the mountains on the horizon, rising effortlessly again each morning with the dawn of a new day.

Activity at the airbase never subsided, but seemed to increase or decrease based on the hours of available daylight. With summer having already arrived upon the interior of Alaska by the middle of June, early morning work schedules were no different than early afternoon or late evening on the airbase. Aircrews, mechanics and support personnel maintained a constant flow of various aircraft to and from destinations all over Alaska, delivering personnel, supplies, mail and the aircraft themselves for the ongoing war effort in the Aleutians Islands and many outlying bases.

Larger numbers of aircraft, both small and large, were arriving on a continuous basis as part of the massive Lend-Lease program with the Soviet Union. Painted with new insignia and outfitted for their continued journey across the Bering Strait, the aircraft were destined for the battlefields on the steppes of Soviet Russia.

A lesser number of daily flights also departed east through Canada along the 1,900 mile airway system of the Northwest Staging Route, returning ferry pilots to the continental U.S. for additional aircraft to be flown north, as well as carrying mail, cargo and official correspondence between the many airfields from Fairbanks to Great Falls, Montana.

The time was 3:30 in the morning as Northwest Airlines captain Gill Enger entered the main hangar at Ladd Field,[1] known as Hangar One on the north side of the runway, and walked up the flight of stairs to the operations and weather office. He was one of many pilots and crewmembers employed by major airline companies, working under contract with the U.S. Army Air Forces' Air Transport Command during World War Two. Assigned to the Alaskan Wing's flight detachment at Edmonton, Alberta, he and the other two experienced men in his crew were part of a large contingent of civilian aircrews operating in Alaska

Parked P-39s in front of Hangar 1 at Ladd Field, Alaska, near Fairbanks. (Blake Smith)

and Canada in support of military operations. Northwest Airlines had been contracted for the route by the military since April 1942 and Gill Enger had been flying the missions almost from the beginning.

Entering the wooden double doors into the main office, he greeted a young Army officer on the way out and nodded to the civilian forecaster standing behind the counter shuffling through a collection of recent weather charts. Having performed the weather briefing routine on many previous occasions, they exchanged pleasantries by first name before going over the synopsis and expected conditions between Fairbanks and Whitehorse in the Yukon Territory of Canada.

---

1 Ladd Field was renamed Fort Wainwright Army Airfield after General Jonathan Wainwright in January 1961.

Forecast weather along the four hundred and eighty-six mile route was generally favorable, calling for scattered to broken clouds from six thousand to nine thousand feet with widely scattered rain showers and unrestricted visibility. Winds were predominately out of the northwest from ten to fifteen miles per hour at flight altitudes between four thousand and eight thousand feet. Light icing was forecast above the freezing level from nine thousand to fifteen thousand.

After receiving the weather briefing Gill Enger filed a visual flight rules flight plan, or contact flight as it was referred to in 1943, intending to follow the air-

1940s view of Ladd Field looking south. Hangar 1 is in the far left of the photo. (Randy Acord)

way system east from Fairbanks over Big Delta, Tanacross, Northway, Snag and Aishihik to Whitehorse. A follow on flight plan would be filed at Whitehorse and other subsequent stops as needed throughout the day. Total time en route for the flight to Whitehorse was estimated at two hours and fifty-five minutes, utilizing a cruising speed of 160 mph. With nine hours of useable fuel aboard, the aircraft could easily remain airborne for another six hours.

Looking up at the sky for the second time that morning as he departed the main hangar, Enger smiled at the patchy overhead clouds only partially shielding the bright glare of the rising sun. In less than an hour he would be flying high above the earth, cruising in clear skies with the full warmth of the summer sun shining through the cockpit window, beckoning him back home to his wife and infant daughter, only a day away. With the forecast for fair weather en route to Whitehorse and even better weather further south, and provided the aircraft could be serviced and cargo transferred in a reasonable amount of time at each stop between Whitehorse and Edmonton, he would be home by evening.

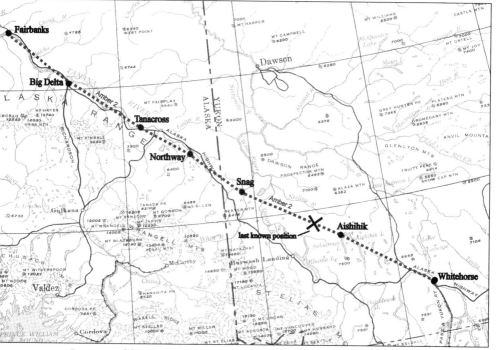

Depiction of the airway route from Fairbanks, Alaska to Whithorse, Yukon Territory.

Several transports and smaller utility aircraft were parked in front of the hangar and along the west ramp as Enger walked toward a C-48 sitting near the runway, being refueled by a green and yellow, dust covered tanker truck. Further away on the east side of the hangar, rows of P-39 fighters and A-20 medium bombers destined for the Soviet Union were visible sitting in a large area of the tarmac. A lesser number of aircraft were parked on the south side of the airfield across from the two parallel runways, near three smaller hangars still under construction. The majority of planes were twin-engine B-25 medium bombers and C-47 transports waiting to be turned over to Soviet pilots for flights across Alaska and the Bering Sea to Siberia.[2]

A large portion of Ladd Field was still under construction in 1943, but the air-

---

[2] From 1942 until the end of World War Two in 1945, nearly 8,000 American manufactured fighters, bombers, trainers and transports were transferred through Ladd Field to the Soviet Union under the Lend-Lease Act.

field was a major hub for military aircraft supporting the war effort. Lend-Lease aircraft destined for the Soviet Union were flown north from the major staging base at Great Falls by American pilots as far as Ladd Field, where they were then turned over to a large contingent of Russian pilots, mechanics and support personnel. Only after the various planes had been inspected by Soviet officials were they accepted and assigned to Russian aircrews for the remaining flight across Alaska to the Soviet Union.

Lend-Lease P-39s at Ladd Field waiting to be painted with Soviet insignia. (Randy Acord)

Co-pilot Raymond Vanderbush waited casually beside the cargo door of the C-48 as Enger approached and gave him the weather outlook for the next several hours. Inside the aircraft, radio operator John Redus, who also served as the navigator and load master, finished securing the small amount of cargo they would be carrying and joined them outside. He handed a small clipboard to Enger, explaining two large mail bags and four official pouches of military correspondence with a combined weight of fifty-nine pounds had been received. It was the extent of their cargo until they reached Whitehorse. There were no passengers.

After briefing the men on the route and contact flight clearance, Enger conducted a general walk-around inspection of the aircraft, duplicating what the co-pilot had done only a short time before. Neither pilot took offense to the other's actions, knowing it was an added safety factor that many pilots performed as a matter of routine. He finished the pre-flight checks as the fuel truck pulled away and verified each of the wing fuel caps was firmly attached before motioning the other members of the crew he was ready. Less than ten minutes later the three men were strapped in their seats with the cockpit checks completed and the engines running smoothly at idle.

A takeoff clearance was received from the tower operator, whose upper torso was clearly visible in the large windowed enclosure protruding from atop the main hangar a short distance away. Enger acknowledged and gave the controller a friendly wave before diverting his attention back inside.

Once both engine temperatures and manifold pressures registered in the normal range, the aircraft was throttled forward until it began moving slowly ahead. After intercepting the painted yellow line of the taxiway it increased speed and continued forward until reaching the eastern end of the pavement, where it turned into position on the runway threshold. Within seconds the lightly loaded C-48 was powering down the new runway in a crescendo of noise, lifting effortlessly into the early morning air as a swirl of gray exhaust smoke and dust settled slowly to the ground. The time was 4:31 am.

Except for the dull green military paint covering the once highly polished metal skin, the twin-engine C-48B looked like any one of hundreds of other Douglas DC-3s and DSTs initially manufactured for the civilian airline industry. Enger's plane was originally operated by United Airlines as a DST-A, which was the Douglas Sleeper Transport version of the popular DC-3A. It was one of three DST-As requisitioned into military service from United Airlines by the U.S. Army Air Forces in March of 1942, then given a new designation as a C-48B.

Both the DST-A and DC-3A were virtually identical in appearance and performance, except for an internal configuration providing sleeping berths for fourteen passengers instead of the more standard seating. The DST-A model had improved 1,200 hp Pratt and Whitney Twin Wasp Radial R-1830 engines instead of the Wright Cyclone engines installed on previous DSTs. It had a maximum range of 2,125 miles, a maximum airspeed of 230 mph, a ceiling of 23,200 feet, and could carry more than three tons of cargo in addition to fuel and crew once the sleeping berths and more expensive accommodations were removed.

Climbing west from Ladd Field over the city of Fairbanks, Enger retracted the landing gear and turned left over the gray, silt-colored Tanana River, heading east toward the small town of Big Delta on a general course following the airway system. At eight thousand feet they passed through the last layer of clouds stretching across the lower landscape, leaving the snow-capped pinnacles of the Alaska Range poking above the pillowed surface to the south and smaller peaks of the White Mountains visible through breaks in the broken cover to the north. Below their flight path the Tanana River and recently completed gravel road known as the Alaska Highway were easily recognizable, running on a similar track paralleling the airway system.

Approximately thirty-five minutes after departing from Ladd Field, Enger's C-48 passed Big Delta and altered course to continue following the airway over the Tanana River and road system toward the Canadian border. Flying above the scattered and broken cloud layers at eight thousand five hundred feet, the tops of the high mountains remained clearly visible less than thirty miles away. In the distance the line of mountains tapered away to the southeast and intercepted the much larger peaks of the Wrangell and St. Elias Mountain Ranges rising skyward for hundreds of miles.

The C-48B, later flown by Gil Engler, is shown to the front of converted DC-3s earlier in 1943 at Edmonton, Alberta. Lend-Lease P-39s are parked on both sides. (Canada Aviation Museum, #7177)

All position reports were normal and on time, as was the next radio call at 5:34 am when passing over Tanacross. Fifteen minutes later the crew sent a transmission to Northway Radio, informing the controller they were over the station and operating as filed under contact flight rules.

East of Northway the terrain began changing in elevation. Forested hills and lower mountains became more predominate after passing into Canada, with several large lakes and river valleys covering the landscape. The highest terrain within a few miles either side of the airway now reached more than six thousand feet, while much further south dozens of ice-encased peaks of the St. Elias Mountains rose more than twice that height. Even though the terrain offered greater risks, especially in marginal weather, as long as an aircraft stayed over the airway or road system it could follow a safe course all the way to Whitehorse.

Routine status reports were again received from the C-48 at 6:20 near Snag

59

and 6:40 am between Snag and Aishihik. Nothing further was heard from the flight and it failed to arrive at Whitehorse as planned. Only forty-five minutes from its scheduled destination, the C-48B and its three-man crew simply vanished, never to be seen again.

In the first few days following the aircraft's disappearance a general area search between Northway and Whitehorse was conducted with military assets, under the assumption the plane had gone done within a short distance of the airway system. When no evidence of the C-48 or crew was found, the search expanded over a wider area utilizing a more systematic grid system to optimize the coverage. Eighteen aircraft flying five hundred and fifty hours over the next four weeks failed to locate any sign of the missing plane. The search was officially abandoned on 13 July.

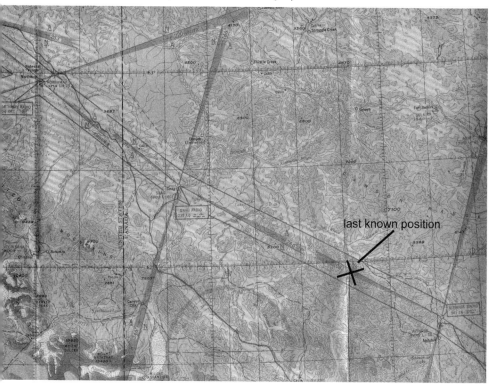

1940s World Aeronautical Chart showing the radio range system between Northway and Aishihik.

There were no reasonable explanations for the sudden disappearance. Weather conditions were later determined to be predominately as forecast during the flight except for a slight change in wind direction near the border from northwest to northeast at fifteen mph, which would only have changed the time en route by a few minutes. Navigation should not have been a factor as visibility was unlimited, and if any reduced visibility was unexpectedly encountered the crew could have easily continued flying on instruments over the airway system.

54th Troop Carrier Squadron C-47s at Shemya, Alaska in 1945. (3rd Wing Historical Office, Elmendorf, AFB)

The circumstances left more questions than answers. It's possible the pilots might have deliberately strayed off the route for a sightseeing tour of the mountains in the immediate area or to divert around one of the isolated rain showers that were mentioned in the forecast, but it is doubtful they would remain off course for more than a few minutes without notifying a controller. Even if they had changed their flight path to proceed direct to Whitehorse, the difference in distance from staying over the airway and road system was insignificant.

US Army Air Forces C-53, a version of the C-47, belonging to the 42nd Troop Carrier Squadron in Alaska. (3rd Wing Historical Office, Elmendorf AFB)

All radio ranges and communication stations along the route were reportedly operating normally during the duration of the flight. Five other military aircraft flying in the area during the same time period experienced no difficulties with navigation or communication. All were operating at altitudes from six thousand to ten thousand feet along the airway system. Radio traffic for hundreds of miles was clearly monitored by other aircraft and ground facilities in the area. Any distress calls or routine transmissions from the missing plane would have been easily received.

The crew of Enger, Vanderbush and Redus were experienced in Alaska and Canada flight operations, having been assigned on the Northwest Staging Route since the previous winter. Both the pilots were instrument qualified with a combined total of almost four thousand hours in the air.

A mechanical failure aboard the missing C-48 was deemed unlikely by the Accident Board since no problems were reported with the aircraft during several routine radio transmissions from the crew over the course of the flight. Even a loss of one engine, if unreported, would not have been critical, since the light payload would have allowed the aircraft to maintain altitude and climb if necessary until reaching a safe landing area.

A catastrophic accident involving structural failure or impact with the terrain would seem to be the most obvious conclusion, but does not explain why wreckage or debris has never been found, especially when the aircraft most likely crashed in proximity to heavily traveled ground and air routes. The Accident Board believed the dark green color scheme painted on the airframe could have precluded search aircraft from observing wreckage in many of the thickly forested areas where the plane might have crashed, but five subsequent decades of human activity in the region have failed to locate any sign of the plane. Numerous low-flying aircraft operate along the same approximate route every year and the area is also well traveled by hunters, trappers, miners and outdoor enthusiasts on a regular basis.

During World War Two many military aircraft crashed in the wilds of Canada and Alaska. Most were found within a few days, some even years later, but a few still remain missing today. Unlike the others, Enger and his crew vanished in fair weather without any forewarning of trouble. Perhaps someday the mystery will be solved.

Chapter Four

February 18, 1944

Caught in the Air

By the early months of 1944, World War Two in Alaska and the North Pacific was going well for American and Canadian forces battling the Japanese. Enemy troops on Attu had been defeated in May of the previous year and other Japanese troops on Kiska evacuated from the island in July, only days ahead of a massive Allied invasion. With their two footholds in the Aleutian Islands of Alaska now lost, Japanese forces in the North Pacific were forced back into a defensive position around their home islands.

Once the Japanese threat was eliminated from Alaska in the summer of 1943, bombing missions by Army and Navy units began against enemy targets in the Kurile Islands of Northern Japan. Newly established American bases at Adak, Amchitka and Shemya, and later at Attu and Kiska, enabled medium and heavy bombers of the Army's 11th Air Force and Navy's Fleet Air Wing Four to strike the Japanese homeland for the first time. Attacks against northern Japanese bases continued until the end of the war, drawing critical enemy air and sea forces away from the main Allied advance further south across the Pacific.

Aircraft combat losses from the beginning of hostilities until the end of the war in the Aleutians were heavy at times, but small in number when compared to the total number of aircraft destroyed. The majority of losses were often attributed to horrendous weather conditions which battered the Aleutian Islands for most of the year, and to a lesser extent the mainland of Alaska. Out of almost five hundred Allied aircraft which were destroyed or seriously damaged in the Alaskan Theater during World War II, only fifty-six were a result of enemy fire. Other aircraft simply vanished, never to be heard from again.

Forty military aircraft were listed as missing in Alaska between June 1942 and December 1945. Some disappeared on combat missions or patrols and others while on routine transport or training flights. Almost all have never been found. One of the worst periods of non-combat losses occurred during a few days in February 1944.

A strong storm system moving across the North Pacific through the Bering Sea, swept along the coast of Alaska and further inland into the interior with devastating results. High winds, low clouds and heavy snow on the perimeter of the storm battered the Aleutian Islands and Gulf of Alaska as the center of the air mass pushed violently inland, gaining strength against the mountains before spilling over into the low river valleys of the Kuskokwim, Tanana, Yukon and Koyukuk. Smaller weather systems influenced by the severe storm shifted direction or stalled, causing their own unpredictability.

One of the first aircraft caught by the fast moving and unexpectedly harsh weather was an obsolete Douglas B-18A "Bolo" bomber with a seven man crew. It departed Shemya Island for Amchitka Island on 16 February, only to vanish in marginal conditions generated near the edge of the tempest pushing across the western Aleutians. Three months later the bomber was found where it had fatally crashed on Kiska Island.

Hundreds of miles away from Kiska Island in the Gulf of Alaska, a Pan American DC-3A operating under contract with the U.S. Navy also disappeared on 16 February near Cape Yakataga. Only small pieces of flotsam and a few human remains were found in the area during the next several days.

By 18 February the horrendous storm which had been building off the Aleutian Islands and Gulf of Alaska hit the southern coast near Anchorage like a runaway freight train. The high winds knocked over telephone poles, uprooted trees and tossed parked aircraft about like bowling pins. Anything not tied down or protected inside a well constructed hangar was seriously damaged or completely destroyed. Winds driving ahead of the storm as far away as Fairbanks damaged many local buildings and homes. Record winds gusting above sixty miles an hour were recorded over a five hour period. In all, the Army Air Forces in Alaska lost thirty-four aircraft to the windstorms on the eighteenth alone, luckily while most were still parked on the ground. A few were tragically caught in the air.

Early in the afternoon of 18 February, a B-25H "Mitchell" Bomber assigned to the Cold Weather Testing Unit at Ladd Field in Fairbanks, left Kodiak with seven occupants en route to Anchorage and Ladd Field. The crew consisted of the pilot, Captain Homer Wilson; co-pilot, First Lieutenant Ray Sharp; gunner, Sergeant Orval Adkins; engineer, Sergeant Clarence Turner; engineer, Corporal Lee Christy; engineer, Private First Class Melvin Rarick, and civilian technician William Marburg.

Chapter 4—Caught in the Air

A formation of 77th Bomb Squadron B-25s off Attu Island in the Aleutians, 1943. (3rd Wing Historical Office, Elmendorf AFB)

A low flying B-25 on patrol near Attu Island. (3rd Wing Historical Office, Elmendorf AFB)

A B-25 from the 77th Bomb Squadron on a raid over the Japanese Kurile Islands. (3rd Wing Historical Office, Elmendorf AFB)

B-25 light-medium bombers were built in large numbers and several variations by the North American Aviation Company for a variety of roles, earning a reputation as the best medium bomber of World War Two. In addition to high and low level bombing, the B-25s were used effectively for photo reconnaissance, anti-ship, submarine patrol and ground attack missions. The B-25H was the most heavily armed of all the variants, with fourteen .50-caliber machine guns, a 75mm cannon mounted in the nose and a bomb payload of 3,200 pounds. A thousand B-25H models were built between 1943 and 1945.

Flying on a visual flight clearance, Captain Wilson climbed through a scattered cloud layer over Kodiak and proceeded on course, unaware of the full intensity of the storm still ahead. Cruising at 230 knots under the power of the aircraft's two Wright R-26000-13 turbo-supercharged radial engines, he headed north toward the mainland of Alaska at 8,000 feet.

B-25s at Ladd Field near Fairbanks being preheated for a winter flight. (Museum of Flight, Seattle, WA)

An hour after takeoff the B-25H reported thirty miles south of Homer and requested an update on the weather at Elmendorf Air Force Base in Anchorage. The current conditions were immediately transmitted back to the aircraft, reporting an overcast ceiling at 1,500 feet with three miles visibility in light snow and strong gusting winds more than thirty knots from the west. Captain Wilson elected to bypass Anchorage and proceed on to Fairbanks where he could get ahead of the frontal system, preferring not to descend through the clouds for an approach and be caught on the ground in the high winds.

The twin-engine medium bomber and its crew were soon flying above a solid overcast extending south of Cook Inlet from the Kenai Peninsula to the Chigmit Mountains of the Alaska Range. As Wilson and his crew neared landfall, strong winds generated by the winter storm pushing over the coast soon began shaking the aircraft with moderate jolts of turbulence.

Intercepting the airway system and climbing to 12,000 feet, the B-25H continued north along Cook Inlet and the wide Susitna River Valley, next reporting forty miles west of Anchorage over Skwentna. Flying in heavy clouds on instru-

B-25H of the 77th Bomb Squadron. Note the 75mm cannon and .50-caliber machine guns mounted in the nose. (3rd Wing Historical Office, Elmendorf AFB)

77th Bomb Squadron B-25H awaiting its next mission in the Aleutians. (3rd Wing Historical Office, Elmendorf AFB)

ments without the aid of visual references, Captain Wilson and his men soon began experiencing severe bouts of turbulence north of Talkeetna. The violent surges of wind battered the aircraft and sent their stomachs in turmoil, at times flinging the powerful twin-engine bomber uncontrollably about the sky.

To make matters worse, severe icing was developing in the heavy, moisture laden air inside the clouds. As the ice accumulated on the wing surfaces, the de-icing system inflated rubber boots on the leading edges, breaking the ice loose. This sequence was repeated every few minutes. Provided the deicing system reacted as rapidly as the ice accumulated, the plane was in no immediate danger.

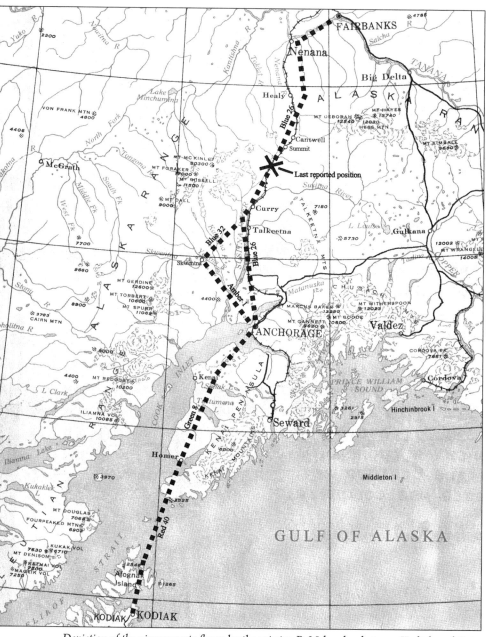

Depiction of the airway route flown by the missing B-25 bomber between Kodiak and Fairbanks in 1944.

Chapter 4—Caught in the Air

Continuing along the airway past the village of Curry toward Broad Pass and Summit Field, Captain Wilson attempted contact with Fairbanks Radio. When no reply was received after two attempts, he broadcast in the blind their estimate for passing Summit Field in ten minutes and arriving at Ladd Field in Fairbanks thirty-one minutes later. Both calls were received by Fairbanks Radio and acknowledged, but the replies were apparently not monitored by Captain Wilson. No further radio transmissions were received from the B-25H. It was never seen or heard from again.

Weather reports from various stations along the airway system at that time all showed extremely strong winds, snow showers, low ceilings and low visibility. When Wilson and his crew should have been passing Summit Field, the base was reporting clouds at the surface, zero visibility and heavy snow with powerful wind gusts. Severe icing and severe turbulence were also prevalent between Talkeetna and Healy.

That same afternoon a P-39 Airacobra and P-51 Mustang were caught unexpectedly by the weather and crashed in nearby Broad Pass, killing both pilots. Both aircraft were attempting to land at Summit Field when they went down. The fighters were reportedly flying low in heavy snow showers. An Army C-47A cargo plane, also caught by the severe weather over the Alaska Peninsula across from Anchorage, crashed killing three of the six man crew.

Over the next several days an extensive search was conducted for the missing B-25H within fifty nautical miles on either side of the airway from Curry to Fairbanks. With no results after several days, the search expanded north of Fairbanks along a fifty mile wide corridor to Fort Yukon, further east of Curry as far as Glennallen, further west to Mt. McKinley and south all the way to Kodiak. Nothing was ever found.

Search conditions were hampered by several feet of fresh snow, which continued falling over the search area in the days immediately following the plane's disappearance. Previous accumulations of snowfall already exited at the lower elevations of the search area and much heavier concentrations of snow in the higher mountains, all of which could have easily hidden any wreckage from view. Even so, every suspicious object, unusual shadow, avalanche and strange pattern in the snow was thoroughly investigated, without success.

The aircraft accident board which investigated the incident believed Captain Wilson continued toward Ladd Field in instrument conditions or attempted to climb above the weather after his last radio contact. Unable to receive a reliable radio signal because of precipitation static, the aircraft was probably blown off course into the higher mountains east or west of the airway, or continued past Fairbanks unaware of its position until running out of fuel. It was estimated that

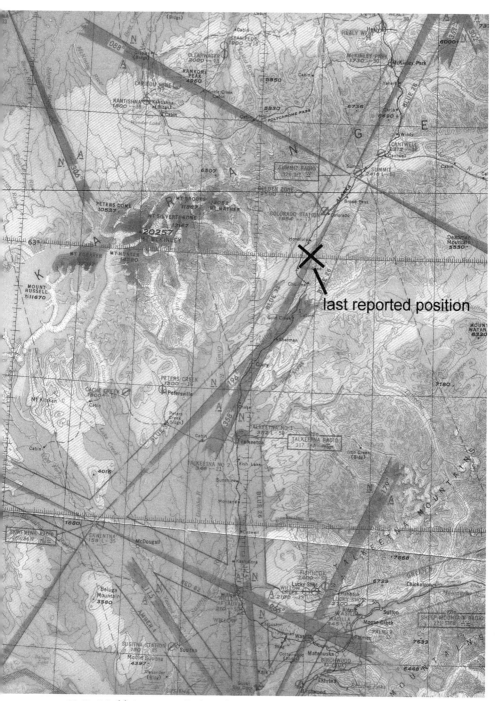

1940s World Aeronautical Chart showing the radio range airway system where the missing B-25H disappeared.

winds aloft were from a southwesterly direction at the time of the disappearance, which could have pushed the bomber at a much faster ground speed than the crew anticipated. Structural failure due to severe turbulence or the inability of the aircraft to maintain flight because of severe icing was also determined to be a strong possibility. But where was the wreckage?

Because no additional radio transmissions were received from the bomber after it passed Curry, it seems likely the crew encountered some sudden and violent episode which made further radio contact impossible. Since no wreckage has ever been found along the airway system in an area well traveled by air and ground traffic for the past sixty years, the aircraft probably went down somewhere among the massive glacial ice fields and towering snow-capped peaks of the Alaska Range.

Without an accurate navigation signal to track the airway because of precipitation static, which was a frequent occurrence with the radio range beacons of that period, stronger than anticipated in-flight winds could have pushed the aircraft east off the airway. Also, if the pilot overcompensated for the winds and applied excessive drift correction, the plane would have been pushed off course to the west. Both directions on either side of the airway have numerous high mountains and remote glaciers that could hide even the largest of aircraft.

Renewed search efforts were conducted the following summer after the winter's snow had melted from all but the highest mountains and glacial valleys, with no result. Six decades of subsequent summers have also failed to reveal any trace of the missing bomber.

Other planes have disappeared in the same area since the B-25H was lost. One was a twin-engine Cessna 410B on a commercial charter flight to Fairbanks in July 1984. Unlike the missing WWII bomber, it was equipped with modern navigational equipment, but also flying on instruments over the airway system north of Talkeetna. It too has never been found.

Chapter Five

March 25, 1944

Empire Express

A cold, biting wind was blowing steadily across the tarmac, chilling even the hardiest members of the five aircrews as they cursed and prepared their twin-engine PV-1 aircraft for another long, strenuous mission across the treacherous North Pacific to Japan. Working carefully around the snow and ice covered ground, each man performed his duty in a ritualistic manner ingrained from months of training and the much more valuable lessons learned during subsequent months of hands-on experience. They were all aware an accident under such hazardous conditions was not uncommon and remained focused on the tasks at hand.

Situated on the eastern side of Attu Island in the Aleutian Island chain of Alaska, the joint Army and Navy airbase at Casco Field was a temporary home for the men of Patrol Bomb Squadron VPB-139 during World War II. Assigned as part of Fleet Air Wing Four to Attu in early December 1943, the squadron crews performed primarily patrol and search missions around the Aleutians for the first month to become familiar with the area and weather. By January the mission expanded to weather and combat reconnaissance, then actual night bombing and night reconnaissance of Japanese installations in the Kurile Islands of northern Japan. Those long over-water missions across the North Pacific became known as the *Empire Express*.

As part of the large scale strategic plan for the assault and invasion of Japan by American forces in the Pacific, the bombing missions from the Aleutians were primarily intended as a diversion in order to draw Japanese forces north, away from primary targets in the southern and central Pacific. By strengthening their bases in the Kurile Islands with increasingly more assets, Japanese forces were spread over a wider area, weakening other bases and strongholds along the primary Allied route of advance.

While flying the dangerous missions from their remote base on Attu, PV-1 aircrews were not only fighting the fanatical Japanese, but almost continuous bad weather and marginal living conditions. High winds, low clouds, reduced visibility in persistent rain and snow, and in-flight icing were as common as the northern Pacific waters were cold. Maintenance, refueling, armament, navigation and communication were all adversely affected in some way by the weather, but the crews endured, kept their spirits up and continued flying in spite of the conditions.

The first PV-1 bombing mission of the Japanese northern islands occurred on 20 January by a flight of three Venturas flying from Attu. Further missions were flown during the next three nights, and subsequent missions during the months that followed. Although the long over-water night flights in severe weather conditions were extremely hazardous, their initial success encouraged larger flights with as many as six PV-1 aircraft. Other Naval air units in the Aleutians flew similar missions, while Army aircraft kept up a bombing campaign with their larger four-engine B-24s and twin-engine B-25s during the day.

Covered with snow and ice, PV-1s wait on their next mission from Attu Island, winter 1944. (George Villasenor)

The five Lockheed PV-1 Venturas departing on 25 March were each fitted with a seemingly small internal load of only three 500 pound bombs, five photoflash bombs and twenty 20 pound anti-personnel fragmentation bombs. This allowed for maximum fuel capacity on the fifteen hundred mile roundtrip to the Japanese Kurile Islands. Once loading was completed the planes were preflighted, the aircrews briefed and the engines run-up in preparation for takeoff. The aircraft were fully refueled only after taxiing into position on the departure end of the runway, thus maximizing the range before accelerating and lifting off into the night sky.

On departure each plane would climb several thousand feet through the thick overcast into hopefully clear skies and then navigate westward toward Cape Lopatka on the southern end of the Kamchatka Peninsula. From there they would turn south for targets at Paramushiro or Shimushu in the enemy con-

Winter preparation for a mission on Attu Island. Removing snow from the wings and preheating were common tasks. (George Villasenor)

PV-1 Ventura being loaded with an anti-ship torpedo on Attu, winter 1944. (George Villasenor)

trolled Kurile Islands. The aircraft did not join together in formation, but flew separately at different altitudes and time intervals. Each plane would also return on its own, averaging almost nine hours of flight time for the entire mission.

Navigation was partially accomplished with low frequency radio signals, when

A flight of Lockheed PV-1s on patrol near Attu Island. (George Villasenor)

the transmissions could be received clearly, which was rare, or by celestial navigation when aircraft could fly above the overcast. Usually weather conditions dictated otherwise, however, requiring the aircraft to navigate by dead reckoning alone, using time, distance and heading until locating a landmass after several hours of flight with the onboard radar system or visually through a break in the clouds. Fuel was always critical, often made even worse by low fog and visibility over Attu and the other Aleutian islands on their return. Ditching at sea was not uncommon and aircraft occasionally disappeared entirely.

The PV-1 Ventura was the military version of the Lockheed Model 18 Lodestar. It first entered service with the USAAF as a medium range coastal patrol bomber, but was later produced primarily for the Navy beginning in December 1942. Initially designated as a B-34 and B-37 with the USAAF, aircraft produced for the Navy were designated as the PV-1 Ventura. Designed primarily for maritime patrol, the Navy PV-1s were modified with additional fuel tanks, increased

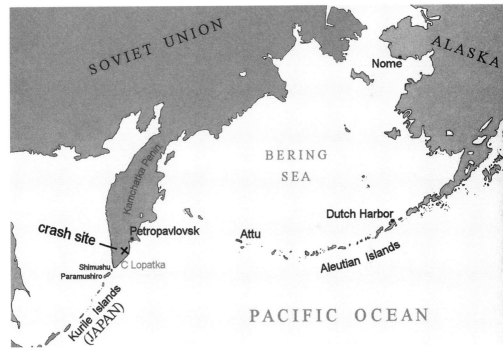

Map of the North Pacific-Alaska, Kamchatka and Japanese Kurile Islands.

bomb loads and search radar. The internal bays were also configured to carry six depth charges or one torpedo, as the mission dictated. Versions of the PV-1 saw service with many Allied countries throughout the war.

The PV-1 Ventura was fast and powerful with two Pratt and Whitney R-2800-31 air-cooled radial engines, capable of 2000 hp for takeoff and a maximum speed of 322 mph; faster than most Japanese fighters produced during the war. Cruise speed was a modest 170 mph, providing a maximum range of 1,660 miles. Armament included two forward firing .50-caliber machine guns in the nose, twin-mounted .50s in an overhead turret and two .30-caliber machine guns in flexible positions near the belly.

On 25 March the first PV-1 aircraft to depart from Attu lifted off just after midnight from runway 10, utilizing most of the 6,700 foot long airstrip. The other four aircrews and ground support personnel on the Navy side of the airfield watched the glow from the wing and tail lights fade into the dark overcast as the drone of the engines lingered above the sound of the wind for a few seconds longer.

Lighting was kept to a minimum on the airbase because of possible enemy attack, leaving only a few low intensity runway lights as a reference against the black night. Above the clouds the moon and stars provided comforting celestial illumination, but below there was only heavy darkness that masked every-

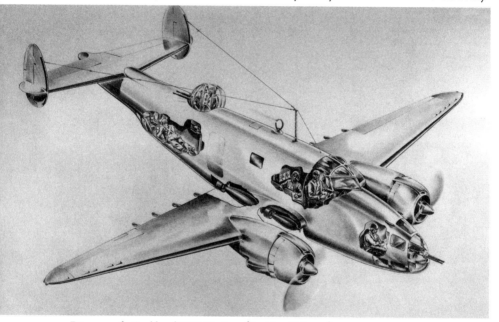

Diagram of a Lockheed PV-1 Ventura (Lockheed Martin)

thing more than a short distant away. Gilbert Ridge, a barren rocky outcropping extending out from the island northeast of the field and the higher Henderson Ridge to the west, were all but invisible. If the crews did not already feel isolated on such a gloomy evening, they soon would be, flying alone over the vast, cold waters of the North Pacific.

Four minutes after the first aircraft departed, the second PV-1 on the mission accelerated forward down the runway in an almost identical takeoff roll. As the plane left the ground it seemed to maintain a level flight altitude for several seconds before beginning an unexpected and gradual descent into the waters of Massacre Bay. The engines were still maintaining power as it skipped hard over the surface and bounced upward, seemingly hanging suspended in the air for a few seconds until falling rapidly into the waves once again. This time the impact was more severe, causing the fuselage to break apart and burst into flames. Only three of the seven man crew escaped before the broken aircraft was consumed by fire and sank. The three shaken survivors were quickly picked up by a rescue boat standing by for just such an emergency.

A PV-1 on Attu Island being refueled before the next mission. (George Villasenor)

Personnel on the airfield watched in shock as the PV-1 descended below the level of the runway, only to hear the explosion and watch the distant flames burning brightly on the surface of the water. They could not know the fate of their fellow airmen, and only hoped they had somehow managed to get clear of the aircraft before it burned. Those thoughts were quickly replaced with youthful determination as they pushed their personal feelings aside and continued with the mission. Canceling the flight was not an option.

Parked and ready, a PV-1 awaits its crew at Casco Field on Attu Island. (George Villasenor)

The third aircraft was airborne after a thirty minute delay waiting for the crash boat to get back into position, and was followed ten minutes later by the fourth PV-1. Both took off normally over the bay as spilled fuel still burned on the surface. In a few seconds they disappeared into the clouds, turning west along the south end of the island for the long flight toward Japan.

Lt Walter Whitman and his six-man crew waited in the last of the five PV-1 Venturas. He and his men were all experienced crewmen, with the exception of a new aerographer on his first combat mission. The crew consisted of the co-pilot, Lt (jg) John Hanlon; Navigator 2nd Class Donald Lewallen; Radioman 3rd Class Samual Crown; Machinists Mate 2nd Class Clarence Fridley; Ordnanceman 3rd Class James Palko; and Aerographer 2nd Class Jack Parlier.

Repeating the takeoff sequence of the other four aircraft, Lt Whitman maneuvered his Ventura across the icy taxiway and into position on the runway, checking each of the instruments and engine indications for normal operation. After shutting down the engines on the takeoff end of the runway, they waited for the fuel truck to pull along side and completely top off the tanks. Once completed the engines were restarted and checked again before preparing for the takeoff roll.

PV-1 crews from Fleet Air Wing Four preparing for a mission. (3rd Wing Historical Office, Elmendorf AFB)

Keeping the aircraft aligned on the runway by applying slight pressure on the brakes, the throttles were advanced until the two powerful Pratt and Whitney R-2800 engines were roaring at maximum takeoff power. The plane accelerated slowly at first, then began increasingly in speed as the hangars, shops and living quarters passed behind out the right cockpit window. A thousand feet from the end of the runway the nose came up, rotating the fuselage into the air only a few hundred feet from the water. Suddenly all outside references were lost as they continued away from the few lights on the airfield over the dark water of Massacre Bay and into the low overcast. Transitioning to instruments, Whitman climbed steadily ahead on a fixed heading until reaching two thousand feet, confirming his altitude before banking right in a continuous climb over the radio range station on the southeast corner of the island. From there they navigated outbound on the southwest leg for a few minutes until turning on an approximate westerly heading toward the Kamchatka Peninsula.

Whitman's PV-1 continued to a higher cruising altitude after breaking through the overcast at seven thousand feet, maintaining a set course off of celestial bearings taken by the navigator and fading signals from the radio station at Attu. Without even those simple references they would have been flying blind, trusting the forecast winds were accurate and would deliver them on course over the distant landfall.

A tight formation of PV-1s on deployment. (Lockheed Martin)

Approximately two hours into the flight their outside visibility was again lost as they entered a higher cloud layer extending well above their cruising altitude. Snow was present in the clouds, causing precipitation static which began interfering with radio reception and blocked any possibility of an accurate bearing from the Soviet radio beacon on Cape Lopatka. For the next hour and a half they would be flying across the open ocean by dead reckoning alone.

PV-1s on a training flight in the Aleutians. (George Villasenor)

Ahead of Lt Whitman's PV-1 the other three mission aircraft were flying under similar circumstances. The first aircraft was an hour ahead and began experiencing problems with its starboard engine while still short of the Kamchatka Peninsula. It began consuming a larger amount of fuel than normal, requiring the plane to abort the mission and return to Attu while it still had reserves for the flight home.

Aircraft number three, which was now the lead aircraft on the mission and a half hour ahead of Lt Whitman's plane, sighted landfall through a break in the overcast after four hours of flight. The PV-1 began experiencing severe icing and severe downdrafts over the peninsula, but continued south by navigating on radar until arriving over the island of Shimushu at 9,000 feet. While still flying in the clouds it was able to deliver its bomb load over the target. After ten hours of flight it finally landed back at Attu with barely enough fuel to taxi off the runway. Winds at their cruise altitude on the return leg were ninety degrees off from what was originally forecast, causing a quartering headwind from the southeast instead of a quartering tailwind from the southwest.

Twenty minutes ahead of Lt Whitman's aircraft, the remaining PV-1 on the mission reached the Kamchatka Peninsula north of Cape Lopatka, but the crew could not get a fix on their exact location. Flying blind in the heavy clouds and experiencing severe icing and updrafts, they were unable to get oriented or locate the intended target. Not wanting to chance dropping their bombs on Soviet territory, the crew turned for home and jettisoned their bombs over the ocean. They too landed at Attu after ten hours of flight with little useable fuel remaining in the tanks.

Lt Whitman and his crew would have been unaware of what the other three aircraft were experiencing ahead of them, but most likely encountered the same weather conditions as they continued toward the Kurile Islands. What actually transpired during the flight is a mystery. The last known contact between Attu and Lt Whitman's PV-1 occurred six hours into the mission when he requested a radio fix on their position. Presumably the aircraft was en route back to Attu at the time. The airbase informed him the transmission showed the plane on an approximate magnetic bearing of 263 degrees from Attu. The reply was acknowledged, but nothing further was heard from the flight.

When the aircraft failed to arrive at Attu by the time its fuel would have been exhausted, an organized sea and air search was conducted over the waters west of Attu. Two destroyers, a destroyer escort and fifteen aircraft extensively searched an area between 240 degrees and 330 degrees from the base in an arc out to a distance of two hundred miles. Nothing was ever found. Speculation arose the plane had simply run out of fuel and ditched somewhere in the harsh North Pacific, where evidence of their fate simply vanished into the depths. Over the passage of time they were all but forgotten.

Fifty-five years later in October 1999, a package arrived at the United States embassy in Moscow containing brief information and pictures of a wrecked twin-engine aircraft with American markings on a remote mountain on the Kamchatka Peninsula. The information also provided the name of a Russian geologist who found several human remains at the crash site during a geological expedition in 1962. A Joint United States-Russia Commission on POWs/MIAs substantiated the information during an interview with the Russian geologist three months later and forwarded their findings to the Central Identification Laboratory in Hawaii for further review.

During the interview in January 2000, the Russian geologist explained how he and his survey team first stumbled upon the site in 1962 and the subsequent discovery of four sets of human remains. He went on to detail the layout of the site and how they removed personal items from each of the bodies. Those items, mostly jewelry and money, were later turned over to a KGB official after reporting their finding of the wreckage.

At the time of the geologist's discovery the fuselage was found facing uphill, relatively intact except for the two engines lying approximately 200 feet below the main wreckage. The geologist also remembered observing several bombs and loose ammunition scattered around the crash site.

The Joint Commission was later able to confirm the Soviet military visited the site during another geological team's survey of the area in 1970. One of the Russians who was with the geological team claimed they found a partial set of remains near the wreckage and another set several hundred yards north of the site in an old streambed. He stated that while they were still in the area a military helicopter arrived with a small group of soldiers, who exploded the live ordnance still lying around the wreckage.

Although the type of aircraft and date of the crash were not known by the Joint Commission, it was believed to be one of several possible missing aircraft which were lost in that area during the war. A team of commission members and specialists from the Central Identification Laboratory were organized for a late summer expedition when the winter snow would be gone from the slopes and the weather allowed easier access. The team hoped to initially determine the type and serial number of the aircraft and whether human remains still existed. If so, an archeological recovery team would be flown in to conduct a detailed analysis.

A mission of such complexity would not even have been considered, much less allowed, before the collapse of the Soviet Union in 1991. With the fall of communism and formation of a democratic government in the newly established Russian Federation, relations between Russia and the United States became much more amicable, allowing access to Soviet records and past intelligence on missing personnel from the Cold War and World War II. Many aircrews lost during those decades are still unaccounted for.

Representatives from both the Russian and American governments were later included in the investigation team, as well as an anthropologist, archivist and technicians from the Central Identification Laboratory. They arrived by helicopter at the remote crash site, located five miles southeast of Mutnovsky Volcano and seventy miles south of Petropavlovsk, the capital city of Kamchatka, on 7 August 2000.

Kamchatka is relatively uninhabited except for a few small cities and military bases, with large areas of isolated wilderness covering much of the 750 mile long peninsula. Two large mountain ranges containing many active volcanoes extend diagonally across the remote landmass. Inside the southern range on a remote alpine covered slope, nine miles from the coast, is where the PV-1 Ventura wreckage was finally located.

As the team first approached the site they could clearly distinguish large pieces

Wreckage of the missing PV-1 found near Mutnovsky Volcano on the Kamchatka Peninsula. (CIL-Hawaii)

of wreckage in an open area of low vegetation and lichen covered rocks. Large areas of darker scrub brush surrounded the wreckage and patches of winter snow were still visible on the higher slopes. Pieces of the aircraft were visibly strewn along a slope in a 50 by 130 yard area, with a torn section of the cabin and wing the largest and most visible. Both engine assemblies were located further down slope, 75 yards behind the main area of wreckage, where they had apparently torn loose during the crash landing.

Upon closer inspection of the wreckage the investigation team located one unexploded 500 pound bomb, numerous .50-caliber rounds and a clearly distinguishable 34641 number stenciled on the tail-section. The number positively identified the aircraft as a Navy PV-1 assigned to VPB-139, reported missing on 25 March 1944 with a crew of seven.

Several pieces of human bone fragments and personal items were recovered inside the broken fuselage by the investigation team. Also found near the wreckage were three distinct craters, confirming the Soviet military had detonated some of the other explosive ordnance carried on the aircraft. The four .50-caliber and two .30-caliber machine guns aboard the aircraft were not found and were assumed to have been removed or destroyed by Soviets troops in 1970.

Scattered wreckage litters the remote mountain slope. (CIL-Hawaii)

No excavation of the crash site was attempted by the team in 2000. Upon completion of their three day investigation the team departed, with a strong recommendation that a complete recovery operation be conducted in the near future.

A second team arrived the following year in early August with enough supplies for four weeks and the proper equipment necessary for a thorough excavation

A tail section and wing are identifiable among the wreckage. (CIL-Hawaii)

of the crash site. The ten-person recovery team from the Central Identification Laboratory spent the next thirty-five days on the Kamchatka Peninsula dismantling and removing every piece of wreckage, searching every inch of ground and excavating the soil as much as a foot below the surface. Every item recovered was carefully photographed and cataloged for later analysis.

Several aircraft components, personal items and bone fragments were removed, while other objects were left at the site. Most items located were in and immediately around the wrecked cabin section of the aircraft. Recovered items included a wallet, pocket knife, eyeglasses, dog tags, coins, tools, boot fragments and parachutes, an altimeter and the aircraft data plate. All the old ordnance at the site, including the 500 pound bomb, ammunition, smaller explosives and flares, were disposed of. Personal effects taken from the bodies in 1962 by the

Distant view of the crash site with Mutnovsky Volcano on the right. (CIL-Hawaii)

Soviet geological expedition could not be confirmed as having been in official custody and were not recovered.

Although the cause of the crash was not determined in the recovery team's investigation, the type of damage and pattern of wreckage observed at the site provided some possible clues. There was apparent battle damage visible on one and possibly both of the engines, which could explain why the aircraft had made an emergency landing. Articles written about the aircraft after it was investigated in 2000 and a television documentary on the subject stated the damage was probably from Japanese anti-aircraft weapons or fighters it encountered near the target. With a crippled aircraft and some of the crew possibly wounded or dead, the pilot might have attempted to reach a neutral landing site at the Soviet airbase in Petropavlovsk. Once safely over the peninsula the aircraft was probably unable to maintain sufficient power and was forced to make an emergency landing.

That assumption could be correct, but it is just as likely Lt. Whitman's crew encountered the same severe weather conditions as the other aircraft on the mission and never found the intended target. In an attempt to identify their location the crew could have strayed too close to a Soviet shore battery and been fired upon by Soviet anti-aircraft artillery. There are several accounts of Soviet forces in that area firing on American aircraft during the war, and aircrews that bailed out or made emergency landings were subsequently interned, along with their planes. Lt Whitman's PV-1 had a complete bomb load on board when it crashed, which indicates it had not reached the target area. Since the load could easily have been dropped over

Each section of the wreckage was sectioned off in a grid pattern. (CIL-Hawaii)

enemy territory or in offshore waters before returning to the peninsula, it is likely the plane was forced to ditch before approaching the Japanese islands.

Six hours into the mission the crew contacted Attu with a request for a radio fix of their position, presumably with the intent of returning back to base. At that time the aircraft had probably not yet been damaged. Only after the aircraft sustained engine damage over the Kamchatka Peninsula did Lt Whitman elect an emergency landing to the nearest suitable piece of terrain he could find, probably by descending through a small break in the overcast. Evidence suggests the plane touched down in a level attitude on a southeast heading, with the engines shutdown or at a low power setting.

There was no indication of any post crash fire and damage observed to the aircraft by the Russian geologist in 1962 appeared survivable. A video taken of the site in 2001 before the excavation began was made available to the Russian geologist, and he stated much of the wreckage appeared to have been disturbed in the decades since he had originally been there.

Since four partial bodies were also found outside the aircraft in 1962, it's possible some of the crew initially survived the crash, only to succumb to the harsh winter conditions or injuries incurred during the landing. Other members of the crew might have attempted reaching a settlement for help, perishing later

One of the investigation team members taping off a section of wreckage. (CIL-Hawaii)

somewhere in the remote and unforgiving mountains. Some human remains were also found inside the wreckage in 2001.

A boot fragment from one of the crew was found 300 yards north of the site in a dry streambed, substantiating the story of the Russian geologist who was at the site in 1970. The name of one of the men on board the PV-1 was clearly stamped on a worn leather name tag. Either the remains were drug to that location by wild animals, which frequented the area, or the individual had sought shelter in the depression on his own before dying.

Three potential grave sites were also found near the wreckage by members of the recovery team, but an excavation did not uncover any human remains. It's possible some of the bodies were buried by surviving crewmen following the crash, or later by Soviet troops in 1970, only to be removed some time in the future.

Evidence suggests at least one or more of the crew either bailed out over Japanese controlled territory or managed to walk out of the mountains after the crash. They were then captured by or turned over to Japanese troops or interned by the Soviets. A television documentary of the crash recovery shown on PBS in 2001, mentioned the notorious Japanese radio personality "Tokyo Rose", who only four days after the plane disappeared claimed it had been shot down. She listed the full names of each of the crewmembers, including the co-pilot's nick-

A view of the main area of wreckage awaiting analysis. (CIL-Hawaii)

name. It is doubtful the Japanese could have obtained such specific information so quickly unless at least one of the men was captured. But was the information obtained by local Japanese troops or relayed by a Soviet contact? And what happened to those men? Were they killed after being captured, or imprisoned only to perish in a Japanese camp or Gulag? If so, why wasn't their fate revealed when the war ended? We will probably never know, unfortunately, as no evidence has ever been found confirming the capture of any of the crew.

There is also little known about the actions of the KGB and Soviet military regarding the missing Ventura and its occupants after it crashed. The fact Soviet troops visited the site at least once is an almost certainty, since statements place soldiers there in 1970 and some ordnance from the plane was destroyed at that time. Whether the Soviet government was aware of the site or visited the site before that date is unknown. A more detailed investigation into Soviet activity at the crash site over the previous five decades was recommended by the recovery team, but there is no indication any further information was discovered or if a further investigation was later attempted.

Remains of the once missing aircrew found at the crash site were transported to the Central Identification Laboratory in Hawaii for analysis once the expedition was completed. DNA evidence was then extracted from samples of the recovered bone fragments and compared to blood samples from surviving relatives of the seven crewmembers. Only two of the crew could be positively identified and those were returned to their families. The other human remains were buried together in Arlington National Cemetery outside Washington D.C. with full military honors.

Chapter Six

November 30, 1945

A Cold Winter Morning

Lieutenant (jg) John McMillan had his jacket collar turned up around his neck as a barrier against the biting, cold wind blowing across the airfield at Kodiak, Alaska. He tried not to hurry and look intimidated by the weather, keeping his hands inside the pockets of his worn leather jacket and cursing himself for leaving his gloves and cap back at the hangar. It had only been a short walk to the Flight Operations Office to check the weather conditions, but halfway across the tarmac he regretted not dressing more appropriately. Now that the wind was hitting him square in the face on his walk back to the aircraft, it felt even colder, numbing his exposed cheeks and causing his eyes to water.

A dark, early morning ceiling of low clouds was masked behind the bright lights of the hangars, maintenance shops and offices on the southern edge of the airfield. The sun was still well below the horizon and the winter moon, illuminating on the few nights it was visible, was once again hidden above the thick blanket of gray. Weather conditions were not expected to improve. Snow flurries were forecast by early afternoon and the winds were forecast to increase to as much as 40 knots. By evening the clouds would be thinning, but the winds would be increasing further, gusting to 50 knots.

It was a good time to be heading home, Lieutenant McMillan thought as he opened the hangar door and stepped inside. The warm air from the powerful overhead heaters hit him immediately, enticing him to open his jacket in relief. In another hour his seven man crew and cargo of seventeen passengers would be safely over the Gulf of Alaska, on their way home to Washington State and their waiting families. For most of the servicemen it was their first trip stateside since the war ended, finally being furloughed in time for Christmas. For others it was a chance at a well deserved rest before returning for duty, or in the case of the flight crew, another routine mission back to waiting wives and girlfriends at Whidby Island Naval Air Station outside Seattle.

A four-engine PB4Y-2 Privateer patrol bomber stretched over half the width of the large hangar as Lt. McMillan entered, completely enclosed from the winter weather inside the painted walls and interlocking sliding doors. Another identical aircraft, also scheduled on the mission, was parked further back at an angle on the opposite side of the hangar. A few mechanics and technicians from the late shift were scattered across the building, finalizing various repairs and inspections.

One of the large aircraft hangars at Kodiak Naval Air Station, with Lockheed P2Vs, a Consolidated PBY-5A and Douglas R4D (DC-3) parked on the tarmac. (USN)

McMillan's crew was thankful for the available space, which allowed their daily maintenance and preflight checks to be conducted in relative comfort before the mission. A third PBY Privateer was parked in an identical hangar nearby, their crew preparing for the same flight across the Gulf of Alaska. Other aircraft were sitting outside on the parking ramp and taxiway, waiting for repairs or their own crews for later missions.

Hangar space was not always available, especially during the winter months, and aircraft were often left outside and preheated using large, portable Herman Nelson style heaters. The main wheeled body was heavy, but attached with light, flexible conduits, which could be easily positioned in separate locations, much like the tentacles of an octopus. The four engine Privateers, because of their large size, usually required a minimum of three heaters; one near each wing for heating the engines and a third to warm the interior cockpit and cabin area.

A PB4Y-2 at Casco Field on Attu Island before its next flight. (George Villasenor)

All three aircraft assigned to the flight were Navy PB4Y-2 Privateers, a derivative of the Army Air Force B-24 Liberator used for high altitude bombing. In outward appearances the B-24 and PB4Y-2 were very similar, except for some minor structural modifications and internal configurations. Most noticeable was the twin or forked tail of the B-24 being replaced with a larger single vertical tail on the Navy PB4Y-2. The Navy version also had a longer nose section enclosing a ball turret, a second top turret behind the wings and side fuselage turrets in place of the open machine gun windows for a heavier defensive capability.

A PB4Y-2 showing its heavy armament. (1000aircraftphotos.com)

Normal armament was twelve .50-caliber machine guns. Additional radio and radar antennas also gave the Privateer an improved search capability. Internally the Privateer contained more electronic systems and crew space than a B-24, as well as additional fuel tanks in the bomb bays necessary for long range patrol. The four Pratt & Whitney engines had also been modified with higher power ratings for lower flight requirements, including replacing the turbo superchargers installed on B-24 Liberator engines with mechanical superchargers. A bomb load of 12,800 pounds could be carried if necessary, although with a decreased fuel range. The aircraft had a top speed of 247 mph at a maximum ceiling of 19,500 feet and a maximum range of 2,900 miles. It carried up to twelve crewmembers while on normal patrol missions, but a lesser number when configured as a transport.

The three Privateer crews were not concerned with long range patrols on November 30, 1945. World War Two was over and each of their aircraft had since been modified to carry additional passengers in place of bombs and armament. The crews were more than happy providing transportation for their fellow servicemen heading home. Most had been busy for months flying passengers back and forth from Kodiak, but seats were still a luxury. One of the lower ranking Navy men on base had been trying for days to get a flight stateside where he could be released from active duty, but available space was always filled with higher priority passengers. Circumstances changed the young man's fate when a Coast Guard Lieutenant overhead his predicament and graciously gave up his own seat, deciding instead to fly on a later commercial flight.

Lt McMillan glanced at his watch, noting the time as 05:20 am as he began briefing the other seven members of his crew on the in-flight weather conditions and their intended route from Kodiak to Seattle, Washington. They would be the first of three PB4Y-2s scheduled to depart, carrying seventeen active duty passengers on the nine hour instrument flight. The other two Privateers would depart at timed intervals behind them, each carrying their own full load of passengers. There were no questions from the crew. Each of them knew their job assignments and could perform each others' tasks almost as well as their own. They were an experienced crew used to long range war patrols over the North Pacific, but still adapting to peacetime duties. The passenger flights were, if anything, a welcome change.

Even though the maintenance and preflight checks had already been completed by other members of the crew, Lt. McMillan conducted his own walk around inspection as the aircraft was being hooked up to the ground handling tug in preparation to being moved outside. Starting forward at the nose and moving slowly along the left fuselage and wing, McMillan paid special attention to the control surfaces, engine intakes and landing gear. In some areas he felt or moved the components with his hands, ensuring the proper movement, tolerance or security, while in other areas he only visually checked the condition. As the large hangar doors slid open away from the nose he continued around the tail of the aircraft, examining the horizontal stabilizer and vertical fin, then repeated the same procedure on the opposite side. Everything appeared normal, as he expected. His PB4Y-2 was a good aircraft and his flight crew and the ground maintenance teams kept it in top condition.

It only took a few minutes before the aircraft was safely repositioned onto the parking ramp in preparation for startup. Lt. McMillan was the last of the crew to climb aboard. The co-pilot, Ensign Joseph Osebold, was already seated and they exchanged disparaging remarks on the weather as he moved into the left pilot's seat. The rest of the crew consisted of radiomen Charles Baker and Herman Bumbus, machinists mates Benjamin Norman and John Spengler, and ordnancemen Haywood Waller and John Plevelich.

A bus pulled up outside and began unloading the passengers as the pilots went through the pre-start checklist. When a ground crewman gave the all clear signal, each engine was started in sequence and checked for normal operation. Engine pressures and temperatures were closely observed for any sign of malfunction. The auxiliary power units used to start the engines were then disengaged and moved a safe distance away, followed by the portable fire extinguishers which were being manned by other ground personnel.

Sounds of the passengers getting situated in the back were drowned out by the engines, but once everyone was secure the flight engineer notified the pilots through the intercom. The engines were then taken through the run-up sequence, again monitored for proper pressures, temperatures and rpms. Navigation systems, radios and onboard radars were checked, flight controls, fuel, hydraulic and electrical systems verified. Each crewmember then called in over the intercom, acknowledging they were secure and each station was operating correctly.

A flight clearance and taxi instructions were requested and received from the ground controller before Lt. McMillan signaled the ground crew to remove the wheel chocks. Once they were pulled safely clear of the plane, Ensign Osebold advanced the throttle controls just enough to slowly move the aircraft forward, allowing them to taxi toward the runway. As the aircraft turned ninety degrees along the taxiway, McMillan could see the Privateer from the other hangar being towed into position on the parking ramp. It was Lt. MacLean's aircraft from VPB-120 (Patrol Bombing Squadron 120), who would be departing second on the mission. The other Privateer still sitting in the first hangar was from his own squadron, VPB-122, and was commanded by Lt. Johnston, who would be in the third aircraft following Lt. MacLean. Each of the aircraft filed flight plans for different altitudes and would depart approximately 30 minutes behind the aircraft ahead.

There was no other traffic that early in the morning. The runways were clear except for the lone four-engine patrol plane moving down the taxiway from the

Lt. McMillan's PB4Y-2 with protective covers over the windshield and canopies. (David Strong via Steve Hawley)

two large hangars near the seaplane ramp on the south side of the airbase. After intercepting runway 36 the aircraft continued forward, twice altering course at runway intersections before reaching the threshold of runway 7, a distance of nearly three miles, and aligned itself for takeoff. The tower controller cleared the Privateer for departure as the four powerful Pratt & Whitney engines were throttled to full rpm. Seconds later the brakes were released and the plane lurched forward, slowly at first, then with increased speed as it raced down the runway and lifted skyward into the dark overcast. The local time was 06:18 in the morning.

The missing PB4Y-2 from VP-122 with unidentified crewmen at Kodiak NAS. (David Strong via Steve Hawley)

Lt. McMillan watched the co-pilot level the plane when the aircraft reached its cruising altitude of 10,500 feet, mentally confirming they were on course toward an intercept with the radio range station at Sitka, 600 miles further east across the Gulf of Alaska. Most of the route would be flown by dead reckoning and it was the most dangerous part of the flight. Not because of enemy forces, but because of the few communication and navigation facilities, lack of safe landing

Chapter 6—A Cold Winter Morning

A PB4Y-2 Privateer on maritime patrol over the Pacific. (USN via flyingtherim.com)

Aerial view of Kodiak NAS, looking west down runway 28. (USN)

areas and the vast expanse of the inhospitable, cold waters of the North Pacific. Only after arriving over Sitka would it get easier. From there they would turn southeast along the many coastal islands, following the airway system to Seattle.

Now flying in clear skies with the moon and stars shining brightly overhead, McMillan and his crew relaxed, enjoying the pristine view. The en route weather wasn't expected to change for more than an hour, once they reached 145 degrees longitude, approximately mid distance between Kodiak and Sitka. It was at that point over the Gulf of Alaska that they would encounter higher cloud formations being generated by a stationary cold front south of their course. Light turbulence

PB4Y-2 Privateer taxiing for departure at Shemya Island. (3rd Wing Historical Office, Elmendorf AFB)

and light rime icing was then expected along with rain and snow showers for the remainder of the flight.

Lt. McMillan's flight progressed without incident until intercepting the Amber 1 airway over Sitka. Communications were normal during the three hours en route from Kodiak. The only change was a climb to a new altitude of 12,000 feet in an attempt to get above turbulence encountered after entering the weather conditions associated with the cold front. The other two Privateers were airborne by then as well, flying a more direct route from Kodiak that intercepted the airway system at Masset in the Queen Charlotte Islands of Canada, 200 miles southeast of Sitka. The time was 09:25 am when Lt. McMillan reported crossing the Sitka Radio range at 12,000 feet, estimating an arrival over Masset at 11:05. He gave no indication of trouble or concerns about the weather.

Chapter 6—A Cold Winter Morning

At approximately the same time, Lt. MacLean in the second Privateer, 150 miles southwest, was encountering thick clouds at his altitude of 14,000 feet and increasing turbulence. A few minutes later his aircraft crossed the distant southwest leg of the Sitka Radio Range and entered moderate icing conditions prevalent in the low temperature and heavy moisture. All four of the engine carburetor air temperatures suddenly dropped to zero, causing carburetor icing to develop and impede the flow of fuel to the engines, resulting in a loss of power. Emergency application of full throttle and fuel mixture to auto rich only momentarily alleviated the problem.

Moments later Lt. MacLean's aircraft luckily flew out of the mass of moisture laden clouds and regained normal power, but only after dropping 2,000 feet in altitude. He continued the flight by avoiding as many of the other cloud formations as possible, maneuvering away from more potential icing conditions. Shortly after regaining full power he monitored a broken transmission from Lt. McMillan in the first Privateer near Sitka. Even over the garbled static the tension in his voice was obvious as he exclaimed his aircraft was at 1,000 feet and descending. Nothing further was heard from Lt. McMillan or his crew. The time of the transmission was approximately 9:30 am.

A half hour after Lt. McMillan's aircraft was overdue at Masset, all airports, radio range stations, navaids and communication outlets along the airway network were notified of the missing aircraft. A continuous vigilance kept the systems operating for an extended period in the hope the aircraft was only lost or delayed and could still make it to a safe haven. After twelve hours had passed and only when the planes fuel endurance would have been exhausted was the communication network returned to normal operation.

While a communication search was being initiated for Lt. McMillan in the first PB4Y-2, problems continued for the other two aircraft as they continued flying southeast. In spite of his best efforts Lt. MacLean again encountered instrument conditions near Spider Island in the Queen Charlotte Islands of British Columbia, and as before began losing power on all four engines as the carburetors iced up. Repeating the same emergency procedure of full power application and auto rich mixture failed to have any affect. The plane began losing altitude rapidly in an unintended descent out of 12,000 feet. Reacting quickly, Lt. MacLean reversed course for a few minutes and then turned south toward an area of clear sky he had previously observed. After ten minutes of teeth clenching tension he was finally able to break out and regain power, continuing a controlled descent to 500 feet. He flew the aircraft at a low altitude for the remainder of the flight, dodging additional rain showers and icing conditions until arriving in Seattle.

The third PB4Y-2 flown by Lt. Johnston was able to maintain its flight altitude of 10,000 feet without any loss of engine power, but it did encounter the same unforecast moderate rime icing conditions near Masset over the Queen Charlotte Islands. It arrived safely in Seattle shortly after Lt. MacLean's aircraft.

An intensive search for Lt. McMillan's missing plane was initiated within hours of its disappearance. Naval aircraft were dispatched from Sitka, Kodiak and Seattle, joining military and civil air units from other bases and communities along a 1,000 mile stretch of coast. All other commercial and military aircraft and ships in the area were notified to be on the lookout for possible survivors. Assets from the Royal Canadian Air Force soon joined the effort as well, unfortunately bad weather conditions along the coastal islands interfered with search efforts on an almost continuous basis

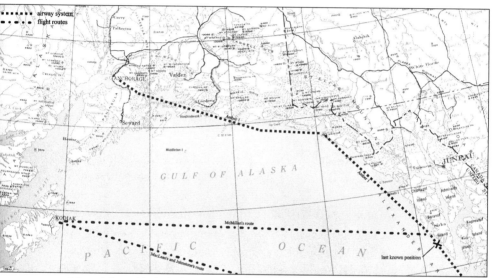

Map of the Gulf of Alaska and coastal area showing the airway systems and PB4Y-2 flight routes from Kodiak.

for days. Rough seas, high winds and low clouds, typical of the coastal islands along Southeast Alaska and Canada in the winter, plagued the region for much of the operation. Even so, a systematic coverage of the area was eventually accomplished by numerous aircraft and ships, although without success.

One promising lead on the missing aircraft materialized the same day it disappeared, when several individuals near Sitka reported seeing smoke late in the morning on 30 November, several miles inland on Baranof Island. If Lt. McMillan had turned his aircraft toward Sitka after losing power, the plane could have crashed in an attempt to find an opening through the overcast. Heavy icing could have caused a malfunction of the onboard radar, rendering the coastline and higher terrain invisible until it was too late.

Following the possible sighting on the island, ground search teams of Army, Navy and civilian personnel were organized, but hampered by high winds and deep snow for several days. One team was able to reach the general area of the smoke sighting on 3 December and reported seeing a dark patch of freshly disturbed snow on the side of Mount Katlian, nine miles north of Sitka. The next day a ground team from the Navy vessel *Hemlock* and two civilian aircraft were able to get within close proximity of the disturbed area, but only found a recent snow slide and no evidence of wreckage from the missing aircraft.

Other ground teams searched the southern shore of Prince of Wales Island, southeast of Sitka, in case the aircraft continued southeast along the airway before it went down. It is the largest island in the archipelago and a likely area the plane might have crashed. A Forty mile expanse of water and a few smaller islands between Baranof and Prince of Wales Island were also covered by aircraft and surface vessels, to no avail.

On 6 December all ground searches were abandoned until evidence of the missing plane could be spotted from the air. Aerial and sea searches were also abandoned within the next few days. Nothing was ever found.

The official Navy investigation into the disappearance of the PB4Y-2 on 30 November remained classified for many years, but was eventually declassified in September 1949. During the post accident investigation a deficiency was found in the design of the R1830-94 engines installed on PB4Y-2 aircraft, which could cause an uncontrolled loss of power during operations in moderate or heavy icing conditions. Immediately after the engine deficiency was identified, all PB4Y-2 aircraft were restricted from further instrument flights in cold weather conditions and no similar accidents occurred. Only after corrective action had been performed to modify the induction system on the engines was the restriction lifted.

It was the Navy's consensus Lt. McMillan's PB4Y-2 encountered icing conditions while operating in instrument conditions shortly after passing the Sitka Radio Range, resulting in corresponding power losses or multiple engine failures due to carburetor icing, similar to that experienced by the second aircraft flown by Lt. MacLean. The accident was not believed to be from any misconduct or incorrect action by the crew.

A review of the maintenance logs determined the aircraft was in good mechanical condition before departure. It had been properly maintained and serviced, was properly configured for transporting passengers and had been fully equipped for instrument flight. No prior deficiencies had been reported on the aircraft or engines since the last maintenance inspection.

Lt. McMillan was regarded as a good, experienced pilot with more than 1,600 flight hours. From statements received by other individuals at Kodiak Naval Air Station, he was observed well rested and cheerful on the morning of the flight. He was fully trained and qualified for instrument conditions and as part of his training had completed special instrument flight training for operations in the North Pacific region.

Both the route and weather hazards were familiar to Lt. McMillan, having flown the same route several times in the past few months. Since the minimum en route altitude between Sitka and Annette was 4,000 feet and Lt. McMillan's last radio transmission stated he was at 1,000 feet, the Navy believed he would not have descended below a safe altitude unless the plane was in serious trouble.

What exactly Lt. McMillan did after the suspected loss of power is purely speculative. It is possible the PB4Y-2 crashed somewhere in the remote mountains of Baranof Island where the wreckage has remained hidden for all these years in the thick canopied rain forest, or perhaps lies on a remote snow covered peak yet to be discovered. In September 1983 the remnants of a U.S. Army Air Forces C-45F Expeditor, the military version of the twin-engine Beechcraft D-18S, were found by accident on a mountain only fifty miles northwest of Sitka on Chichagof Island. The plane and its four occupants had disappeared almost four decades earlier in April 1946.

It is also just as likely Lt. McMillan turned out over the ocean, away from the high coastal island when his aircraft began losing power, in the hope of descending below the cloud cover. There are more than twenty mountains above 2,000 feet in elevation on either side of the airway between Sitka and the southern tip of Baranof Island alone, all of which Lt. McMillan was presumably familiar with and would have tried to avoid. Once below the cloud base in visual conditions, he would have then flown back toward shore in an attempt to make an emergency landing. Either too low or too far away from land when the aircraft broke out of the clouds, the crew was probably forced to ditch in the cold, weather tossed seas, with little chance of survival.

Assuming McMillan's last radio call was accurate, transmitted five minutes after passing the Sitka Radio Range, and his aircraft continued flying at normal cruise speed with an estimated forty knot quartering tailwind, then at the time of the transmission his plane would have been within approximately seventeen miles of Sitka. If, however, the airspeed was slowed just above stall speed to arrest the rate of descent during a loss of power, which would be normal, then the aircraft would have gone down within approximately eleven miles of Sitka. But because the actual time of Lt. McMillan's last transmission could not be verified, the area where his Privateer could have crashed widens considerably with each minute it continued flying.

One has to wonder what must have been going through the minds of the twenty-five crewmembers and passengers as their large four-engine aircraft began descending toward the inhospitable waters along the Alaska coast. Only a short flight distance away from the island community of Sitka, they probably assumed a rescue would be quickly forthcoming, with only the inconvenience of a few short days delay before continuing to their destination. Perhaps a few suspected the worst, but none could imagine they would be only one of many civilian and military aircraft to disappear in the area over the next five decades.

No wreckage was ever found of the missing aircraft, either washed ashore on the thousands of miles of shoreline within the coastal islands, or snagged off the bottom of the sea by the hundreds of commercial and private fishing boats operating in those waters every year. The disappearance of the PB4Y-2 Privateer with twenty-four occupants is a mystery that has endured the passage of time, but only one of many which still exist in the annals of aviation history.

Chapter Seven

February 24, 1947

Weather Reconnaissance

As the glow from the winter sunrise began filtering over the eastern mountains, a lone B-29 Superfortress lifted off the runway from Elmendorf Air Force Base outside Anchorage, Alaska, heading west toward the Bering Sea. The local time was eight o'clock in the morning.

Forty-five minutes later, north of Augustine Island and approaching Bruin Pass in the Chigmit Mountain Range at an altitude of 2,000 feet, the aircraft's radio operator sent the plane's first and last position report. The four-engine bomber and its thirteen man crew were never heard from again.

As far as the public knew the B-29 was on a routine training flight over western Alaska and its offshore waters. Officially the flight was designated as a sixteen and a half hour weather reconnaissance mission, but its true intention was monitoring much more than storm patterns and weather data. Although it was indeed gathering some atmospheric information, the main purpose of the flight was to record radio traffic and unusual activity being conducted at Soviet facilities and military bases in the Kamchatka Peninsula and eastern Siberia.

A Cold War had been rapidly escalating between the Soviet Union and the United States since the end of World War Two. By 1947 the once former ally had become a dangerous adversary in a deadly game of global influence and power mongering that would last nearly five decades. While most politicians did little but debate social issues, ideologies and their own self-importance, it was the U.S. military which remained a true deterrent to communist expansion. Because of Alaska's close proximity to the Soviet Union and many of their military facilities, Alaska became strategically important as a base for protecting American interests in the Pacific. A strong military presence in Alaska, especially from the U.S.

Air Force, served a critical role during the coldest period of the Cold War; one that continues on a lesser, but no less important worldwide scale today.

In 1947 the B-29 Superfortress was still the world's premier heavy bomber. Only four years had passed since the revolutionary new aircraft first entered service in World War Two with American Air Forces in the Pacific. It was the first high altitude strategic bomber designed with pressurized crew compartments and an automatic fire control system. The B-29 could also fly higher, faster, over a longer distance and with a bigger payload than any other bomber of its generation. By the end of World War Two new radar and electronic countermeasure equipment was added, while later versions were configured for weather reconnaissance, photographic missions and aerial refueling operations.

WB-29 from the Weather Reconnaissance Squadron at Ladd Field, Fairbanks, Alaska. (354th Wing Historical Office, Eielson AFB)

More than 2,500 B-29 aircraft were built through 1945, with another 1,400 B-29As, Bs, Cs and Ds entering service with minor improvements. Later configurations were designated as F-13s, RB-29s and B-50s. The aircraft were capable of nearly a 400 mph maximum airspeed, a combat ceiling more than 36,000 feet, a range of more than 5,000 miles and an internal bomb load of 20,000 pounds. A formidable defensive armament included twelve .50-cal machine guns and one 20mm cannon.

The B-29's design was so advanced when it entered service in late 1943 the Soviet Union tried every means of espionage and diplomacy to obtain the technology. The opportunity materialized much quicker and in a much easier manner than anyone imagined when several of the aircraft were delivered right to their doorstep. Toward the end of WWII, three B-29s battle damaged during bombing raids on Japan made emergency landings in the Soviet Union, resulting in the aircrews being interned and the aircraft being systematically dismantled. Within three years facto-

Chapter 7—Weather Reconnaissance

ries in the Soviet Union were constructing exact duplicates for their own Air Force.

In spite of all the B-29's advanced capabilities, today it is still most readily remembered as the aircraft which dropped atomic bombs on the Japanese cities of Hiroshima and Nagasaki in August 1945, effectively ending WWII. The Soviet version of the B-29, known as the Tupolev Tu-4, also gave the Soviet military a means of delivering atomic weapons after exploding their first nuclear device in 1949.

Aircraft 44-86253, a weather reconnaissance B-29-95-BW in the Alaska Air Command and assigned to the 717th Bomb Squadron of the 28th Bomb Group (VH), departed Elmendorf AFB on 24 February on a routine reconnaissance flight. Sometime during the mission the plane and crew disappeared without a trace.

The bomber first entered service on 15 May, 1945. It was a relatively low time

B-29 "Klondike Kutey" at Elmendorf AFB, Anchorage, Alaska. (3rd Wing Historical Office, Elmendorf AFB)

aircraft when it vanished, having flown only a total of 147 hours since leaving the assembly line. It was in good shape for a B-29 and all four of its 2,200 hp Wright R-3350-57 air-cooled, turbo-supercharged engines had been replaced or overhauled within the preceding year.

The aircraft was commanded by Captain Raymond Tutton, an experienced pilot with more than 2,400 flight hours, including 450 combat and 260 instrument hours. His crew consisted of seven other officers and five enlisted men; including pilots Second Lieutenant James Lomax and First Lieutenant William Johnson; navigators Captain Kenneth Frentz and Second Lieutenant Hubert Jarvis; flight engineer First Lieutenant Warren Truhe; bombardier First Lieutenant John Ludacka; electronics officer First Lieutenant Paul Riddle; crewchief Master Sergeant Leon Blakeman; radio operator Staff Sergeant Robert McLaughlin; and gunners Staff Sergeant Clarence Adams, Staff Sergeant Orville Ransom and Staff Sergeant James Maloney.

The planned sixteen and a half hour mission was supposed to take them through Bruin Pass on the west side of Cook Inlet, across the Alaska Peninsula to Naknek, west along the coast to Nunivak Island, south over the Bering Sea, north into Norton Sound, then back south along the coast to Nunivak Island before returning to Elmendorf AFB. Twenty-two hours of useable fuel was carried aboard the aircraft.

Weather conditions for the route were forecast VFR over Cook Inlet, but expected down to as low as one mile visibility in rain and snow showers with a ceiling of 1,500 feet from Bruin Pass south to Umnak Island. A low pressure area centered off the Aleutians was causing overcast conditions and precipitation for most of the remaining route. Moderate to heavy turbulence and icing in clouds was expected up to 11,000 feet.

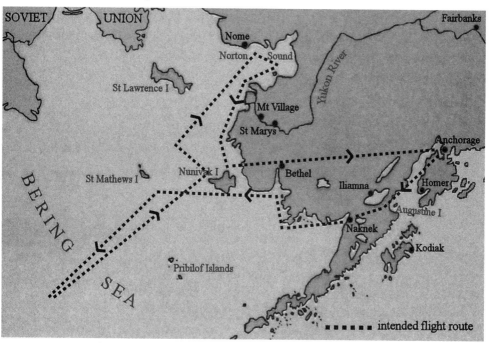

Map of Alaska showing the intended reconnaissance route of the missing B-29.

Since the B-29 was on a classified mission over international waters, radio transmissions were to be kept at an absolute minimum except in the case of an emergency. The Alaska Air Command assumed the sixteen hour flight continued as planned until the aircraft failed to report after exceeding its estimated time of arrival back at Elmendorf AFB. Even then the matter wasn't given much concern, since the bomber had five and half hours of fuel remaining and was assumed to be delayed by in-flight weather conditions. Because no distress calls or position reports had been received from the B-29, radio problems were suspected as a more likely explanation for the crew not extending the flight plan. Only when the overdue aircraft's useable fuel would have expired at 6:00 the following morning, was an emergency alert issued to all military facilities.

Available military aircraft from Elmendorf AFB, Ladd Field, Adak, Kodiak and Ft. Randall were immediately put on alert, but inclement weather conditions over much of southwestern Alaska hindered in-depth search efforts the first day.[1] Only one plane was able to launch from Elmendorf AFB and it was forced to return after three hours. In spite of the demanding effort still ahead and thousands of square miles to be searched, a hope of finding the missing men alive was pervasive among the search crews. It was a positive attitude that was amplified by the report from an individual in Homer, on the south end of the

The missing B-29 on an earlier mission, winter 1946/1947. (3rd Wing Historical Office, Col Alan Hardy Collection)

Kenai Peninsula, who claimed to have seen flares over Cook Inlet on 25 February, the morning the bomber was overdue. Any enthusiasm in finding the missing men soon disappeared, however, when further investigation of the sighting found no validity to the claim. No further sightings were reported in that area.

Military communication sites and civilian stations in Alaska who were notified of the missing aircraft attempted contacting the B-29 bomber with negative results. But there were two likely transmissions from the plane that were later believed to have been received by the Elmendorf AFB tower controller on the morning it disappeared. Tower personnel verified they had received two possible transmissions from the aircraft at 4:30 and 4:50 that morning, more than an hour before the B-29's

---

1 Ladd Field, which was later renamed Fort Wainwright, is located in Fairbanks. Ft. Randall was located at Cold Bay on the southwest end of the Alaska Peninsula, before being deactivated.

fuel would have been exhausted. Although both messages were weak and the first was unreadable, the second transmission stated "6253 calling Elmendorf Airways," before fading out completely. The second radio transmission was also monitored by a different military operator, although only the number "53" was heard before radio contact was lost. Search crews now had confirmation the missing men could still be alive.

B-29 taxiing off the runway after landing at Shemya airbase in the Aleutians, 1945. (Tony Suarez)

Improved weather conditions on 26 February allowed twenty-six Army and Navy aircraft to begin flying over the missing B-29's flight route. The operation included seven planes from the 10th Rescue Squadron, which was in overall command of the search, ten from the 28th Bomb Group at Elmendorf AFB, five from Ladd Field, one each from Adak and Ft. Randall and two from Kodiak. By the end of the day none had spotted any sign of the lost bomber, but some portions of the route remained uncovered. Bruin Pass, the location through the mountains on the western side of Cook Inlet where the bomber would have transitioned through, was not effectively searched until 27 February, having minimal visibility and low ceilings for two days in a row.

Newly arrived B-29 at Shemya being inspected by ground personnel. (Tony Suarez)

B-29 being refueled at Shemya, 1945. (Tony Suarez)

While search aircraft were airborne on the second day of 26 February, the military Direction Finding station at Elmendorf AFB monitored several repeating "V" and "T" Morse code signals from an unidentified source, placing the transmissions northwest of Elmendorf AFB on an approximate 300 degree true bearing. Since the signals were possibly being sent by the missing aircrew, aircraft were directed into that area in an attempt to home in on the transmission. Before any of them could get close enough and lock in a signal, the transmissions ceased. A continued search of the area could find nothing conclusive. One of the aircraft did locate a dark, apparently burned patch of snow northwest of Iliamna, which appeared to be a potential crash site, but it was later determined to be unrelated.

Since it was unknown whether the missing aircraft went down on land or in the sea, even an approximate location could not be determined until some evidence of wreckage or survivors were found. If the B-29 had gone down on land the crew would have had a much better chance of survival, even in the winter. The wreckage would be in a fixed location that could more readily be spotted from the air, survivors would have a better chance of signaling passing aircraft, which were more frequent over land, and if possible survivors could even walk out of the crash area if all other options failed. Shelters could be constructed to provide warmth and protection from the cold weather, and more food and water sources would be available.

At sea, the survivors would be at the mercy of strong ocean currents and winds pushing them away from the location of ditching. The size of the search area would thus be much larger and the number of aircraft involved reduced to only multi-engine planes with adequate over-water survival gear. There would be no shelter available other than the survival rafts, food and drinking water could not be replaced as easily as on land once the rations expired, and storms could abruptly destroy what little they had.

Even so the missing crew was adequately prepared to survive for at least ten days in either situation. Two six-man survival rafts were carried on board the B-29 Superfortress, as well as exposure suits, sixteen cases of emergency rations and water, fourteen sleeping bags, snowshoes, shovels, axes, thirteen first-aid kits, two "Gibson Girl" survival radios, "sterno" fuel cans, two large and eight smaller emergency kits, two flare pistols and a case of flares, and four smoke bombs. Individual equipment included parachutes, a Mae West life preserver, cold weather clothing, a flashlight, snow goggles and personal items. Of course all the emergency equipment was useless if the crew did not survive or sustained extensive injuries.

Poor weather conditions again kept many of the search aircraft grounded on 27 February, delaying an adequate check of the area between Iliamna and Naknek. Confirmation the missing B-29 at least made it through Bruin Pass was received the same day, when a Radio Range operator at Iliamna reported hearing a large aircraft fly over the station in the direction of Naknek at 9:10 on 24 February, the approximate time it should have been in that location. The operator did not establish radio contact. A resident of the village also confirmed seeing a four-engine aircraft flying overhead through the broken cloud cover at approximately the same time. Military and civilian agencies verified no other large aircraft were in the area during that period.

Higher clouds and improved visibility on 28 February allowed sixteen aircraft to again participate in the search, primarily over southwest Alaska between Iliamna and Naknek, and a wide area over the Bering Sea. Bruin Pass in the Chigmit Mountains between Cook Inlet and Lake Iliamna was also effectively covered for the first time. A promising new lead was also reported the same day when a large break in the ice of a frozen lake was spotted by a search aircraft west of Iliamna. That too proved to be a false lead when ground teams and a helicopter dispatched to the lake determined it was a natural occurrence.

As so often happens during a prolonged operation when crews and aircraft are stretched beyond their normal endurance, disaster reared its ugly head and struck another plane involved in the search. Only three days after the bomber disappeared, a sister ship experienced multiple in-flight engine fires before crashing and exploding eighteen miles northwest of Naknek village. Luckily the crew had sufficient time to bail out, but one of the crewmen was killed from head

B-29 from the Cold Weather Detachment at Ladd Field, taxiing at Shemya airbase, 1945. (3rd Wing Historical Office, Elmendorf AFB)

injuries sustained after high winds pushed him into a rocky moraine. His body was finally located after four days. The thirteen survivors were found and rescued within twenty-four hours by three local bush pilots.

Over the next several days the search for the missing B-29 expanded further offshore to include Norton Sound, St. Lawrence Island, St. Mathews Island and the Pribilof Islands. Other aircraft began concentrating south of Norton Sound on the Yukon River near Mountain Village, where residents had observed smoke signals in the distance after a search aircraft flew over the area on 1 March. A local trapper also claimed an aircraft crashed south of St Marys where he had come across a freshly burned area near the river. St Marys is only twelve miles up river from Mountain Village. An intensive search of those locations revealed no evidence of wreckage or survivors. All further searches across the extensive areas of coverage were unsuccessful.

The search for the missing B-29 and its thirteen crewmembers was officially halted by the 28th Bomb Group on 10 March, but the 10th Rescue Squadron, responsible for overall control, continued searching through 25 March, without success. During the extent of the search effort lasting nearly a month, inclement weather routinely interfered with flight operations. Even so the missing bomber's proposed fight route was extensively covered on multiple occasions by several aircraft. A total of 887 hours were flown through 25 March by military aircraft involved in the search, and every reported sighting or possible radio transmission was investigated. Aircraft continued searching in conjunction with routine missions and training flights for months following the official search, in the hope some of the missing men could have survived, but no evidence of the aircraft or crew was ever found.

As in most cases of missing aircraft, unusual reports and sightings often materialize during the investigation that distract from more realistic search efforts. Even so, every bit of information is taken seriously until thoroughly investigated

58th Weather Reconnaissance Squadron WB-50D, a later version of the B-29, with jet assist pods mounted under the wings. (3rd Wing Historical Office, Dave Menard Collection)

or disproved. Sometimes the clues are worthwhile and sometimes not. What at first often seems to be a promising lead usually turns into something entirely unrelated or unsubstantiated. Several such incidents happened during this search.

One such case involved the sighting of aircraft wreckage northwest of Anchorage on Mt. Susitna, which later turned out to be a known crash site from years before. Another involved several individuals from a village south of Homer, who reported seeing a four-engine aircraft flying low under the clouds and heading south on the morning the B-29 departed from Elmendorf AFB. The only problem was they claimed they saw the plane at 3:00, five hours before the missing B-29 even took off. Eskimos from a village on the Yukon River, north of Bethel, said they watched a large aircraft flying at a low altitude to the northwest on 23 February, then enter a steep dive before hearing the sound of a loud explosion. The date they claimed they saw the plane was a full day before the B-29 went missing. Eight different people east of Anchorage also reported seeing a strange blue light in the mountains south of Turnagain Arm. An investigating aircraft later determined it was only the moon reflecting off a patch of ice.

Although many unconfirmed reports can appear fabricated or overly exaggerated when analyzed individually, they become much more important when taken as a whole. In the case of the missing B-29, various sightings from one area seem particularly relevant, even though no wreckage or bodies were ever found. The Mountain Village area on the Yukon River, north of Bethel, had four separate reported incidents that possibly placed the B-29 in that area. Unidentified radio signals picked-up on 26 February by a military Direction Finding station showed three transmissions coming from an area northwest of Elmendorf AFB, on an approximate line with Mountain Village. Natives who reported a large aircraft crashing in the same area on 23 February might have been confused about the date, and the local trapper who claimed a plane had crashed near the river after noticing freshly burnt ground was certain of the evidence. Other residents of the village observed smoke in an area shortly after a search plane flew over, as if someone was trying to signal the passing aircraft.

B-29 at Shemya airbase with the bomb bay doors open, 1945. (Tony Suarez)

Pilots of B-29 #38 conducting a preflight inspection before departing Shemya. (Tony Suarez)

Weak radio transmissions broadcasting the B-29's identification number were heard two hours after it was expected back at Elmendorf AFB and an hour and a half before its fuel would have been exhausted, indicating the plane was still airborne at that time. Since the aircraft was apparently attempting to contact Elmendorf AFB controllers and not another station, it would seem to indicate a close proximity to Anchorage. On the other hand the weak radio signals suggest the plane was either far away or at a low enough altitude that its signal was being partially masked by higher terrain. In the absence of a distress call being received, it would also seem likely that a crash would have been without warning or in an area where radio contact could not be established. It could also suggest the B-29 was experiencing internal communication problems or unusual weather phenomenon, such as precipitation static, that were interfering with onboard radio functions.

Unforecast weather and mechanical problems might have been a factor, but there is no indication the aircraft experienced any difficulty with either during the flight. There was no history of significant mechanical or electronic problems, but the weather was worse than predicted with severe icing and severe turbulence along the Alaska Peninsula and Cook Inlet, as confirmed by another aircraft flying between Naknek and Elmendorf on the same day. The Wright air-cooled engines installed on the missing B-29 also had a history of overheating and catching fire, as experienced by previous B-29 aircraft and another B-29 involved in the search before it crashed. But even if the same situation occurred on the missing bomber, there should have been ample time to transmit a distress call.

There is also a chance the big four-engine bomber strayed into Soviet airspace and was shot down, but as unfriendly as relations were between the Soviet Union and the United States at the time, Soviet documents would still exist indicating if such an encounter occurred. Unfortunately, declassified information from the Russian government since the collapse of the Soviet Union indicates no such incident ever happened.

Although there is a possibility the lost bomber went down in the cold, unforgiving waters of the Bering Sea, lost for all time, it is just as likely the wreckage might be located some day in a remote area of Alaska. The plane could very well be out there now, hidden from view by nature's camouflage, waiting to reveal itself to anyone who stumbles across it.

Chapter Eight

August 6, 1947

Edge of the Storm

Winds were light from the northeast around Kodiak in the early morning, barely disturbing the surface of the harbor in front of the small coastal town. Slow ocean swells rolling in from the southeast appeared to be moving sluggishly through the water, with barley enough force to break against the rocky beaches and outcroppings of the outer islands. A bright sun rising above the horizon was shining through the pale blue sky. The high scattered clouds reflected off the calm waters of the bay and further inland green forested hills stood unmoving against the landscape. Fishing boats of all shapes and sizes were heading out of the harbor toward the productive fishing grounds among the islands and more distant offshore waters. Flocks of seagulls were swooping and diving in a crescendo of noise among the wakes of the vessels, searching and fighting each other for any scrap of fish or garbage thrown overboard by the busy deckhands.

At the Naval Air Station on the south side of the city, a lone Navy PBY-5A was being prepared for departure on a flight to Dutch Harbor in the Aleutian Islands. On board were a five man crew and a full complement of fifteen passengers. En route weather was expected to be predominately visual flight conditions with some lower clouds and visibility further west, but not enough to significantly affect the flight. A total of five hours flying time was estimated along the six hundred mile route.

The PBY-5A lifted off from Kodiak at 06:26 am without incident and turned southwest, climbing above the inland mountains toward Uyak Bay on the opposite side of the island. Flying over the many inlets and bays along the western coast, small silhouettes of fishing boats could be seen stringing out nets to catch the returning runs of salmon, while other boats were busy unloading their catch or waiting to dock at the large, bustling cannery on the south side of Larsen Bay.

Continuing over open water under relatively clear skies, the Navy PBY left the shore of Kodiak Island behind near Middle Cape. Flying across the rougher waters of Shelikof Strait until reaching the mainland, it then turned south along the eastern side of the Alaska Peninsula.

Normal position reports were transmitted every thirty minutes over the next hour and a half. As expected, thirty knot winds from the south were encountered along the Alaska Peninsula, and lower clouds associated with the edge of a frontal system moving in from the Aleutians formed a solid overcast extending across the mountains north of Cold Bay.

Near Sand Point, ninety miles northwest of Cold Bay, low cloud conditions from the surface up to several thousand feet forced the PBY to climb above the

PBY Catalina at Kodiak NAS with wheels still chocked and engines running. (Kodiak Military History Museum, Gehres Collection)

worsening weather while attempting to maintain visual contact with the coast through breaks in the overcast. A radio message passed by the pilot at 10:00 am reported their position in the vicinity of Sand Point, on course at 4,000 feet between cloud layers. If necessary the crew also had the use of onboard radar for identifying their position. The system had been thoroughly checked before departure and could be easily operated if the weather deteriorated below visual flight conditions.

As the flight continued south the weather became significantly worse than forecast until the PBY was completely engulfed in a thick overcast, obscuring all outside references. Flight was now by instruments alone, requiring the pilots to identify their position solely by using navigational aids and the airborne radar. Radio range stations used for instrument flight existed at Ft. Glenn on Umnak

Island and Ft. Randall at Cold Bay, but there were no radio aids in place at Dutch Harbor. A non-directional radio beacon at Ft. Glenn and radio direction finder at Ft. Randall were also available, along with two non-directional marine radio beacons in operation at Cape Sarichef and Scotch Cap on the south end of Unimak Island.

By this time, obviously concerned about the changing weather, the pilot contacted the airfield at Dutch Harbor for current conditions before deciding to continue on. The report sounded promising. Visibility and ceiling were stated as significantly better than the weather at other bases and more than adequate for a visual approach on arrival at Dutch Harbor.

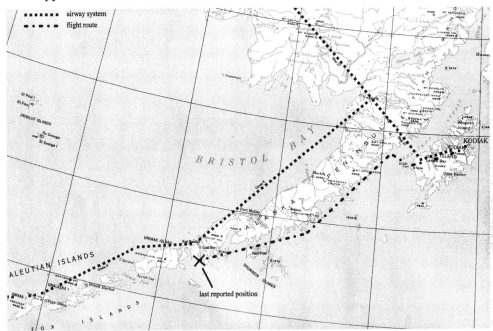

Alaska Peninsula and Kodiak Island, showing airway route and flight path of the missing PBY-5A.

A few seconds after contacting Dutch Harbor the aircraft's radio operator transmitted a position report to the airway controller at Ft. Randall, confirming they were flying on instruments and crossing the southeast leg of the Ft. Randall Radio Range at 4,000 feet. A one hour estimate was given to the next reporting point at Scotch Cap, ninety miles away. The aircraft and crew were never heard from again. An extensive sea and air search over the next several weeks failed to find any trace of the missing PBY or its twenty occupants.

The day began as any other routine mission for the crew. An early wakeup was followed by breakfast before arriving at the airfield to begin preparations for the day's flight. Both pilots entered the Airfield Operations building for a weather briefing and to file a flight plan, while the other three crew members walked out to the ungainly looking PBY-5A sitting on the parking ramp.

A fuel truck pulled alongside the big twin-engine amphibian shortly after the three crewmen arrived and in a short time added enough fuel for ten hours of flight, more than enough for the scheduled mission to Dutch Harbor. Once the refueling was completed the crew inspected the aircraft inside and out, ensuring every component and system was in good working order. The aircraft was the best maintained PBY on the base. It had to be. It was also the designated air-sea rescue aircraft, and as such was kept at a high state of readiness.

A PBY-5A sits on the runway at Kodiak NAS on a cloudy winter day. (Kodiak Military History Museum, Gehres Collection)

Only the day prior the aircraft had undergone a thorough maintenance inspection. Even the most minor deficiencies which had accumulated since the previous maintenance check were fixed or the parts replaced, until the Squadron Engineering Officer considered the aircraft's condition better than at any time in the preceding three months.

Consolidated PBY Catalinas first entered service with the Navy in 1936. Designated as patrol-bomber flying boats, they quickly established a reputation as one of the most durable aircraft in military service. Although the airframe appeared cumbersome compared with many of today's modern aircraft, the unique design was a drastic improvement over previous flying boats, allowing a far greater range and loading capacity. The PBY quickly became popular as a patrol-bomber with many foreign countries as well, serving in virtually every major sea battle during World War II. In addition to its primary role as a patrol bomber, the PBY became famous for night attack missions and as a premier search and rescue aircraft. More than 4,000 variants of the aircraft were eventually built and a few of the planes remain in civilian service today.

Chapter 8—Edge of the Storm

Early models of the PBY flying boat were restricted to only landing and taking off from water, but in 1939 a new amphibious version with a retractable undercarriage began arriving at the patrol squadrons, allowing operations on land or water. Designated the PBY-5A, it became the most widely produced of all PBY aircraft. With two 1,200 hp Pratt and Whitney R-1800-92 Twin Wasp radial piston engines, the PBY-5A had a maximum airspeed of 180 mph, a ceiling of almost 15,000 feet and the capability to carry more than seven tons of weight. Standard flight operations were conducted at a cruise airspeed of 117 mph, permitting a range of more than 2,300 nautical miles.

Aerial view of Kodiak NAS, 1950. (Kodiak Military History Museum, Gehres Collection)

While the enlisted crew of John Duval, Pervis Bangert and Bryce Herndon were busy getting the aircraft prepared for the mission, Lieutenant (jg) William Ziegler, the plane's pilot-in-command, and his co-pilot LT (jg) Nave Fuleihan were going over weather charts with the Navy forecaster. Two strong weather fronts, one near Adak and the other near Ft. Randall were briefed to the pilots, explaining the in-flight conditions they could expect along the route. In addition to multiple cloud layers from the surface up to 6,000 feet along the edge of the storm near Ft. Randall, en route winds were forecast to increase up to thirty knots out of the south. Dutch Harbor was expected to remain above visual flight conditions with a 2,200 foot overcast, light rain and good visibility, while their alternate of Ft. Randall was forecast to be overcast at 8,000 feet with light drizzle. Ft. Glenn on Umnak Island, sixty miles south of Dutch Harbor, had already reported conditions well below approach minimums and was not expected to improve throughout the day.

Fleet Air Wing Four PBY-5A over a winter landscape in the Aleutians. (3rd Wing Historical Office, Elmendorf AFB)

Lieutenant Ziegler decided on a southerly route paralleling the Alaska Peninsula that would take them past Unimak Island, before turning west through Unimak Pass to approach Dutch Harbor from the north. This was the normal route used by other PBY aircrews, since flying through the pass in marginal weather could be accomplished with the plane's onboard radar system. Both Ziegler and the forecaster believed the proposed route allowed the aircraft to fly through the edge of the storm for a shorter period, instead of a more hazardous route across the northern peninsula.

Once the pilots received the weather briefing they filed a flight plan and joined the rest of the crew at the aircraft. Shortly after arriving the pilots completed the cockpit checks, started the engines and checked the aircraft systems, keeping the engines at idle until the passengers and baggage were safely loaded. The fifteen passengers included thirteen members of the Dutch Harbor Army/Navy softball team, who were returning home after participating in an Alaska wide tournament at Kodiak, and two other Navy personnel taking advantage of available seats.

Once airborne the flight proceeded as filed, making standard position reports with no indication of any problems other than a change in weather. Its last transmission was sent four hours and nineteen minutes after departing Kodiak. There were no distress calls, no requests for assistance in fixing their position, and no indications the aircraft experienced difficulties of any kind before it vanished.

PBY-5A on patrol, showing the wheels retracted inside the fuselage. (1000aircraft-photos.com)

When Lieutenant Ziegler's PBY-5A failed to make its next position report and did not arrive at Dutch Harbor in the allotted time, a general alert was sent to all military and civilian stations in the area. Unfortunately there was little they could do except attempt contact on various frequencies, due to low ceilings and visibility from the strong frontal system shutting down flight operations at several of the bases.

A landing PBY slinging mud at an unidentified Aleutian airbase. (Kodiak Military History Museum, Gehres Collection)

The field at Ft. Randall in Cold Bay had been below minimums for most of the morning, much different from what was forecast, and those conditions continued into the next day. A low overcast completely blanketed the surrounding hills and only reached a height of 200 feet over the nearby shoreline. Accompanied by rain and fog, the weather system was generating twenty-five knot winds and reducing the visibility below a mile. Conditions at Ft. Glenn were even worse, with thick clouds hanging to the ground and visibility as low as 1/16 of a mile. Flight conditions were now becoming extremely dangerous as turbulence and icing in the two colliding weather fronts increased in intensity. Dutch Harbor had the only flyable conditions of any station within a couple hundred miles of the weather system, but aircraft could not venture far from the airfield before being forced to return back by an almost impenetrable wall of rain, wind and clouds.

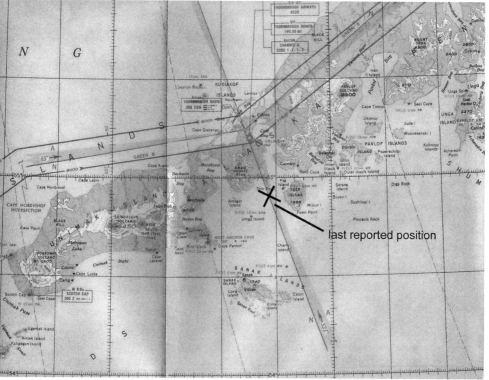

1940s World Aeronautical Chart showing Cold Bay Radio Range and airway system.

A search attempt the first day was primarily left to aircraft from Kodiak, and Ft. Richardson outside of Anchorage, but even those were hampered by severe weather conditions associated with the frontal system moving across the southern Alaska Peninsula.

One aircraft from Ft. Richardson, a B-17 configured for search and rescue operations, did reach Unimak Pass on the south side of Unimak Island three hours after the PBY disappeared, only to encounter a violent onslaught of severe turbulence that threatened to tear the big four-engine plane apart. The pilot managed to get the aircraft turned around and out of the area, but was unable to continue the search. He reported a heavy cloud cover from the surface up to 10,000 feet on the south side of the island, with only scattered cloud layers on the north side.

Several Navy and Coast Guard vessels were also dispatched to participate in the search, but they were forced to struggle through turbulent sea conditions for most of the day just to reach the last known position of the aircraft. Once in the area sixteen foot waves and strong winds throwing off sheets of water combined with low hanging clouds and rain, significantly reducing any chance of spotting potential survivors or debris.

PBY Catalina on a search mission over the Alaska coast. (Museum of Flight, Seattle, WA)

Not until the following day on 8 August was a concentrated air and sea search initiated. Weather conditions along much of the missing PBY's flight route remained marginal, but improved enough to allow two aircraft each from Ft. Randall and Adak to join other planes from Kodiak and Ft. Richardson. Eight surface ships and four submarines were in the area by then as well, combing the shorelines and offshore waters from the Alaska Peninsula to Unalaska Island. In addition, commercial vessels and aircraft operating in Aleutians were also asked to keep a lookout for evidence of the missing plane.

Speculation arose among many of the searchers that the PBY had probably only experienced radio problems and would be found safe once the weather lifted. After encountering the unforecast severe weather conditions, it was surmised the crew probably attempted to land along one of the coastal islands in a sheltered bay or inlet to wait out the storm. Although that scenario was a possibility, as time passed the chance the plane had landed safely seemed unlikely. Hope still persisted survivors would be found, especially since the PBY was well equipped with adequate survival gear for almost any survival situation, but that hope diminished with each passing day.

Over the next week every available aircraft and vessel conducted extensive search operations of the area, often struggling with high winds and low ceilings

An active volcano on the Alaska Peninsula extends skyward behind a PBY Catalina. (Kodiak Military History Museum, Gehres Collection)

that frequented the region. Even so, every possible stretch of shoreline, inland valley, coastal plain and mountain near the plane's flight path was covered as diligently as possible, including several peaks above 8,000 feet.

Shortly after the search began, a promising sighting of some PBY type wreckage was spotted from the air near Dutch Harbor at the 900 foot level on Unalaska Island, but it turned out to be the weathered remains of an old plane from World War II. Another potential sighting of the missing PBY occurred when a submarine noticed aircraft wreckage on a beach near the plane's last reported position, but it too was later identified as an old crash site by a landing party from one of the Coast Guard vessels involved in the search.

Aircraft involved in the extensive operation were unfortunately not immune from accidents of their own. On 10 August a PBY was forced to ditch near St. George Island in the Pribilof Islands, where the occupants were successfully rescued by a nearby Navy ship. Another aircrew wasn't nearly as lucky when their

B-17 crashed five days later while attempting a landing in heavy fog at Cold Bay. All eight Army crewmen were killed.

With no evidence of the lost plane or its occupants located by 15 August, the search by surface vessels was called off and the ships returned to their normal duties. The air search continued for another week, but it too failed in finding any sign of the missing PBY.

The Aircraft Accident Board which investigated the disappearance could not make a determination on the cause, due to a lack of evidence. They did conclude, however, that the weather conditions briefed to the pilots were significantly dif-

Flying in non-typical clear skies along the rugged coastline of the Aleutian Islands. (George Villasenor)

ferent than those which existed near the edge of the frontal system. At the time of the plane's arrival on the south side of Unimak Island, the coastal area was being subjected to extremely hazardous flying conditions being generated by two slow moving frontal systems overtaking each other. This concentration of cold and warm air in the center of the low pressure system caused severe turbulence, precipitation, possible icing and winds from forty to fifty knots, much worse than had been forecast.

The Board believed the possibility of engine trouble was unlikely, given the fact there were several communication outlets in the area who would have received an accompanying distress call and the high level of maintenance sustained on the aircraft. Since no indication of trouble was received from the plane, it was assumed some sudden catastrophic event must have occurred.

Both PBY pilots were considered well qualified and experienced. Lieutenants (jg) Ziegler and Fuleihan each held a valid instrument rating and were familiar with Alaska flying conditions.

Secured with tiedowns against the strong Aleutian winds, a PBY is cleared of snow for the next mission. (George Villasenor)

The disappearance of the PBY became a mystery which has lingered for decades. Whether it crashed unexpectedly into a mountain, broke apart in turbulent air or ditched into the sea, might never be known. Many aircraft have disappeared in the same area and dozens of others still litter the many volcanic islands and southwestern coast of Alaska. Perhaps one day a low flying helicopter, hunting party or hiker will stumble upon what remains of the wreckage, revealing its fate. Until then, the mystery will endure.

Chapter Nine

November 3, 1948

A Routine Training Flight

Small puddles of water collecting on the wet runway were caught by the large spinning tires of the aircraft and sprayed outward, mixing with the light rain that had been falling over the air station for several hours. As the four-engine plane accelerated forward with increasing speed and rotated into the air, the spray of water turned to mist, forming a thin trail of vapor that curled behind the fuselage. The Navy patrol bomber climbed effortlessly into the morning sky, turning southwest away from the air station as streaks of sunlight filtering through the moist clouds reflected off the metal wings. Slowly the sound of its engines and dark silhouette of the fuselage faded above the forested hills, lost against a backdrop of airfield noise and distant clouds.

After departing its home base at Kodiak Naval Air Station on a routine navigation and electronic surveillance training flight, the PB4Y-2 and its twelve-man crew of two officers and ten enlisted men followed the Blue 27 airway across Kodiak Island and Shelikof Strait, flying visually but in radio contact with the airway controller. Cleared for a VFR flight direct to St. Paul Island in the Bering Sea, then back to Kodiak via Cold Bay and Port Heiden, the plane and crew reached the Alaska Peninsula before continuing west off the airway through a pass in the Aleutian Range. Once reaching the waters of Bristol Bay the pilots descended to a patrol altitude of 1,500 feet to begin an electronic radar sweep of the route.

PB4Y-2 Privateers were designed as land based patrol planes during World War II. Modified from B-24 Liberator airframes, the new design added more crew space, electronic equipment, armament and a large single tail fin instead of the twin tail found on B-24s. Since patrol operations were also being conducted at lower altitudes, reconfigured 1,350 hp Pratt and Whitney R-1830-94 Twin

Wasp radial engines were added with higher rated two-speed mechanical superchargers. At 1,500 feet and a cruise speed of 150 knots the aircraft had a range of nearly 2,000 nautical miles.

Piloted by Lieutenant Paul Barker and Ensign Harold Herndon, the long range patrol bomber from squadron VP-20 at Kodiak NAS departed at 7:57 am on the ten hour training flight. The rest of the crew was composed of Midshipman William Musgrove, Chief Aviation Electronics Man William Coleman, Chief Aviation Machinist's Mates Robert Trenton and Franklin Barden, Aviation Machinist's Mate 2nd class Norman Holland, Aviation Machinist's Mate 3rd class Joseph Somers, Chief Aviation Ordnanceman Robert Eichorn, Aviation Ordnanceman 2nd class Lloyd Askildson, Aviation Ordnanceman 3rd class William Clark, and Seaman apprentice James Wooley.

Fleet Air Wing Four PB4Y-2 at unknown Alaska location. Note the large nose turret and single tail, differentiating the aircraft from its B-24 counterpart. (3rd Wing Historical Office, Elmendorf AFB)

En route weather conditions from the Aleutians to the Pribilof Islands were forecast as generally favorable, due to a high pressure area over the Bering Sea generating high broken clouds and good visibility. Much worse conditions from an area of low pressure moving in from the northwest were expected between the Alaska Peninsula and Kodiak on their return, but not of any extreme nature to seriously affect the flight. Routine position reports from the crew to airway controllers were made until 11:30, and from then on hourly coded reports were sent directly to the Navy base at Kodiak.

After arriving at the northern most Pribilof Island of St. Paul in the Bering Sea, 390 nautical miles west of Port Heiden, the PB4Y-2 circled the barren, windswept island inhabited by only a small population of Aleuts and headed back east toward the Alaska Peninsula. Still flying at a 1,500 foot patrol altitude, the cold blue water of the Bering Sea was clearly visible, dotted with small white-capped waves breaking on top of the larger swells from the strong northwest wind.

Two hours into the flight a hazy silhouette of the distant peninsula became visible on the horizon, then darker and more prominent as the range diminished. At first a heavy layer of clouds hanging a few thousand feet above the surface could barely be distinguished from the higher hills rising above the sea, mixing easily with the almost identical shades of land and water. Only as the distance narrowed and the line of thicker clouds became noticeable against the lighter colored sky, and contrasting streaks of dirt and rock and vegetation became visible against the snow covered hills, was the illusion of a single mass of nothingness finally broken.

As the aircraft continued underneath the spreading overcast a few miles out from land, the pilots could see the weather had changed for the worse since their transition through the area earlier that morning. The thick clouds had already closed in around the few passes and valleys along the western mountains and snow was beginning to fall in increasing intensity.

Stretching in a solid line along the Alaska Peninsula, the thick clouds had bunched against the rugged volcanic mountain range rising above the shoreline, obscuring all but the lowest foothills only a few miles inland. Stronger winds gusting to forty knots near the coast also began battering the aircraft with moderate jolts of turbulence, sending anything left unsecured bouncing about the cockpit and cabin.

The Aleutian Islands and Alaska Peninsula were well known for their unpredictable weather. Freak "Williwaw" winds blowing off the mountains and thick sea fog that often materialized out of nowhere had claimed many aircraft over the years. Pilots who took the weather for granted usually paid a heavy price in damaged equipment or loss of life.

At 3:05 pm the crew of the PB4Y-2 Privateer made a routine position and weather report, stating they were approaching the coast at a location forty miles southwest of Port Moller near the south end of Nelson Lagoon, on a heading of 089 degrees. The base of the cloud cover was estimated at 3,000 feet and in-flight winds from the northwest at approximately 35 knots. It was the last communication received from the aircraft.

Exactly what happened to Lieutenant Barker and his crew after their last communication is unknown, but most likely they turned north along the coast in an attempt to cross further up the peninsula and intercept the Blue 27 airway leading into Kodiak. Probably deterred by much worse weather conditions further north, they eventually reversed course and proceeded back to Port Moller in an attempt to follow the inlet waters southeastward across a relatively narrow stretch of the Alaska Peninsula.

A caretaker at the Port Moller cannery on the northwest corner of the bay witnessed a four-engine aircraft flying below the clouds in heavy snow showers at approximately 4:00 pm, heading southeast along the shoreline. A short time later a local trapper ten miles southeast of the cannery observed a four-engine aircraft flying over his cabin, maintaining the same southeast heading. Although neither of the individuals could positively identify the type of aircraft, both were certain it had four engines. It would later be learned no other aircraft other than the Navy PB4Y-2 were operating in the same area at the time.

When Lieutenant Barker's aircraft failed to send any additional position re-

Southwest Alaska map showing the airway system and flight route of the missing PB4Y-2.

ports during the four hours following its last communication at 3:05, a request was issued by the Kodiak Naval District to all stations along the Aleutians and Alaska Peninsula for any information on the whereabouts of the missing plane. After nothing was reported on the aircraft over the next eighty minutes, and the crew's estimated time of arrival back in Kodiak had passed, an emergency warning order was transmitted to Alaska military units to prepare for possible search and rescue operations.

During the extent of the flight after departing Kodiak until the crew's last communication seven hours later, there had been no reports of difficulty with the aircraft or weather.

Even though the plane carried enough fuel to remain airborne until 8:00 pm that night, it was apparent from the lack of communication with the aircraft for such a long period that something serious had probably occurred. Organizing a search to find survivors became an immediate priority.

Weather reports later obtained from stations in the area of the plane's disappearance showed conditions worse than had been originally forecast. Cold Bay, 75 miles south of where the aircraft's last communication was sent, experienced a 700 foot ceiling, five miles visibility in snow showers and northwest winds gusting to 41 knots. Existing conditions all along the Alaska Peninsula worsened significantly throughout day. By late afternoon broken to overcast cloud conditions were predominate from 1,000 to 8,000 feet with occasional moderate snow showers, minimal visibility, moderate turbulence from strong northwest winds gusting to 40 knots, and moderate icing in clouds. Even higher gusts were occurring in the mountains as the winds were funneled violently off the slopes and through the narrow passes, generating severe turbulence that could shake an aircraft uncontrollably.

By 7:49 pm that evening the previous missing aircraft warning order was changed to a specific request for all available military planes at Adak, Elmendorf and Cold Bay to be assigned to a search operation, commencing at first light the next morning. Darkness had already fallen over the area by the time of the request and combined with the poor weather conditions made a search at night both impractical and dangerous.

A highly trained para-rescue team from the 10th Rescue Squadron at Elmendorf Air Force Base was also alerted for deployment to Cold Bay in preparation for a rescue of possible survivors, if and when any were found. Additional notifications were initiated through the federal aviation system to inform civilian aircraft of the missing Navy plane.

Three B-17 aircraft configured for search and rescue missions from Cold Bay were the first to launch on 4 November. They were later joined by six Coast Guard and Navy planes flying out of Kodiak on a wide sweep around the missing patrol bomber's last reported position. When no evidence of the PB4Y-2 Privateer had been sighted by early afternoon, four additional Navy aircraft were deployed from Adak.

The Navy ship *Cocopa*, on patrol in the Bering Sea, was also diverted to search the western side of the Alaska Peninsula between Unimak Island and Naknek, and the Coast Guard cutter *Bittersweet* was dispatched from Kodiak to begin a search between Cold Bay and Katmai Bay on the eastern side of the peninsula. Both areas were remote and relatively uninhabited stretches of coastline extending for hundreds of miles.

Nothing was found the first day indicating what might have happened to the missing plane; not an oil slick, piece of flotsam, recent avalanche or blackened patch of snow. There were likewise no distress calls, flares or other emergency signals identified by aircraft or surface vessels. It seemed as if the plane had simply and mysteriously vanished without a trace.

As the first day of the search operation drew to a close and the hope of quickly finding the missing aircraft faded, the need for additional planes and ships became obvious. Unfortunately, two other aircraft disappearances on 4 November soon

Aerial view looking down runway 18 at Kodiak NAS. (USN)

stretched the search and rescue assets of every available unit from Alaska to the west coast of the United States. One of the aircraft was a Pacific Alaska Air Express DC-3 with seventeen total passengers and crew aboard, lost somewhere between Yakutat and Sitka.[1] The other, a Navy P2V-2 Neptune patrol bomber with nine

---

1 The circumstances surrounding the missing Pacific Alaska Air Express DC-3 are discussed in detail in chapter Ten.

crewmen aboard, vanished during a training flight off the Washington coast near Vancouver Island. More than forty military and civilian aircraft and numerous surface vessels would eventually be involved in the three ongoing search missions.

On the following day of 5 November, fourteen aircraft from Adak, Cold Bay and Kodiak were involved in the search for Lieutenant Barker and his eleven-man crew. Because of the long distance involved in flying between Adak and the search area, the four search planes dispatched from Adak were temporarily assigned to Cold Bay so their additional flight time could be better utilized. Improved weather conditions allowed all fourteen aircraft to effectively search eighty percent of the designated location encompassing more than 15,000 square miles of the Alaska Peninsula.

PB4Y-2 Privateers of Patrol Squadron VPB-122 at Kodiak NAS, secured against the weather. (David Strong via Steve Hawley)

Also on 5 November a report from the Port Moller Radio operator mentioned the possible sighting of a four-engine plane by a cannery worker during the afternoon of 3 November. That report and the subsequent report from a local trapper were later verified and given strong credibility.

Sailors from the Navy ship *Cocopa* spent the next two days on foot and in small boats combing the shoreline around the waters of Port Moller but found no evidence of the missing patrol plane. The ship then headed north and continued its sweep of the outer coast, with the same results.

Poor weather conditions again moved into the Alaska Peninsula and Aleutian Islands on 6 November, effectively halting any significant search attempts from the air. Two planes did manage to takeoff from Kodiak NAS, but both were forced back after a short duration due to low ceilings and visibility. Another two aircraft from Cold Bay also launched, but they were not in the air for any substantial length of time and their efforts were limited to searching a few isolated pockets of coastline not blanketed by heavy fog.

Over the next two days the situation improved dramatically as a high pressure area moved in from the Bering Sea, opening the skies to excellent flying weather. Clear conditions over the northern Alaska Peninsula on 7 November pushed even further south as far as Cold Bay by the following day, allowing multiple flights over the entire search area for the first time. Nine aircraft from Kodiak were able to accomplish one hundred percent coverage in the northwestern sectors, and with the exception of a small tract east of Port Moller, the remaining sectors received fifty percent coverage by other aircraft. On 8 November all the designated search sectors except the mountains around Cold Bay were completely covered by twelve participating aircraft, and the Cold Bay area overall received three quarters coverage.

Underside view of a PB4Y-2 Privateer, with one of its large side turrets clearly visible. (USN)

Good weather never lasts long on the Alaska Peninsula, especially during the winter, and by 9 November all the aircraft at Kodiak and Cold Bay were grounded by terrible flying conditions over the entire search area. But just as quickly as the bad weather moved in, it changed again, providing excellent flying conditions for the following day.

Thirteen aircraft conducted a search from Unimak Island to Naknek on 10 November and except for a few mountains that were surrounded by high clouds, all grid sectors were effectively covered. With no evidence of the missing aircraft or crew located by the end of the day, the formal search of the Alaska Peninsula

PB4Y-2 of Patrol Squadron VPB-111 sits alone at the end of the tarmac. (Louis Bresciano, SV-6, USNR)

area was officially completed. Multiple flights by aircraft over land and water and searches along the coast by surface ships never found a single clue to the missing aircraft's disappearance.

As the detailed search of the Alaska Peninsula and surrounding waters drew to a close, new reports of possible sightings and undecipherable radio transmissions extended the search into other locations. Several residents at Old Harbor on the southeast side of Kodiak Island came forward to report flares had

PB4Y-2 of Patrol Squadron VPB-109 over an inland mountain range. (USN)

been observed further inland only a few days after the plane went missing, and a strange radio signal picked-up and plotted by several surface ships showed a general location approximately 500 miles south of Cold Bay. A second possible emergency radio signal was later plotted almost 200 miles south of the Shumagin Islands east of Port Moller.

Each new incident was taken seriously and thoroughly investigated. Planes began searching the area extensively around Old Harbor over the next two days,

Missing crew members beside their PB4Y-2 patrol bomber, five days before disappearing on a routine mission. Back row-left to right: AOC Robert Eichorn, AO2 Lloyd Askildson, AL1 William Nevares (replaced by ALC William Coleman on the missing flight), SA James Wooley, AD3 Joseph Somers, AO2 Milton Russell (replaced by AO3 William Clark on the missing flight), AD2 Norman Holland, ADC Franklin Borden. Front-row left to right: ADC Robert Trenton, ENS Harold Herndon, LT Paul Barker, MIDN William Musgrove. (Jason Parsons)

but did not locate any signs of an aircraft or persons in distress. Four Navy aircraft returning to Adak from Cold Bay on 11 November searched the area of the radio signal south of Cold Bay, with no results, and additional aircraft covered the waters south of the Shumagin Islands where the second signal was plotted. Both signals were later believed to have come from commercial vessels operating in the area, as the missing Navy PB4Y-2 carried a "Gibson Girl" survival radio which could only transmit on a much lower frequency.

Chapter 9—A Routine Training Flight

The search for Lieutenant Barker and his crew was finally terminated on 12 November after nine days of extensive search efforts over a vast area. Low flying aircraft and surface vessels systematically conducted a visual sweep of the Alaska Peninsula from Naknek south to Unalaska Island in the Aleutians, and from the waters of Bristol Bay to the eastern side of Kodiak Island.

Nearly 500 hours of flight were logged by various aircraft during the search. Although the coverage was thorough and the designated sectors were examined several times, the terrain above 2,000 feet had accumulations of snow from twelve to sixteen feet, while many of the narrow valleys and canyons averaged depths more than twenty feet, all easily capable of hiding a large aircraft. Continuing snowfall during the days and weeks after the plane disappeared could have hidden any signs of wreckage almost indefinitely.

A PB4Y-2 of Patrol Squadron VPB-122 surrounded by snow berms at Kodiak NAS. (David Strong via Steve Hawley)

What actually happened to the PB4Y-2 patrol plane is still a mystery. Whether it crashed or ditched on a remote mountain or into the sea near its last known position might never be known. Many of the mountains in the vicinity of its last known position are higher than the altitude it was flying and some extend well above 8,000 feet. The waters on either side of the Alaska Peninsula are also notoriously dangerous, especially during poor weather that existed at the time of the disappearance. Rough seas and strong winds would have quickly dispersed and concealed any signs of wreckage that had not already broken apart and sank.

It's also possible the crew continued flying for some time after their last position report and crashed much further away while searching for a route around the weather. Icing, turbulence or a loss of visual flight in low clouds and visibility were probably contributing factors. Dozens of mountain passes, high peaks, lakes and inlets scattered across the thousands of square miles of terrain on the Alaska Peninsula can hide even the largest aircraft from view. Vaster expanses of ocean on either side of the peninsula could conceal an aircraft for eternity. Only time will tell the true story, if at all.

Both pilots were experienced and well qualified for the routine training flight. Lieutenant Barker held a special instrument rating and had accumulated more than 3,300 total flight hours, almost half in PB4Y-2 aircraft. He had also flown more than seventy hours of instrument flight in the preceding six months. Ensign Herndon possessed nearly 800 hours of flight time, 480 of those hours in PB4Y-2s.

The aircraft was considered in good condition by maintenance personnel and an investigation by the Accident Board came to the same conclusion. Maintenance records showed the plane had received a ninety hour inspection less than three weeks before it disappeared and had been completely overhauled only seven months previously. The next major maintenance inspection was due eighteen hours from its time of departure.

After the search for the missing PB4Y-2 was terminated a Notice to Airmen (Notam) was issued and later included in the Alaska's Airman Guide, advising aircraft operating between Naknek and Cold Bay to be on the lookout for the missing plane. Less than a month later, a Northwest Airlines flight operating on the airway southwest of Naknek spotted aircraft wreckage on a windblown mountain ridge, only twenty-five miles from the coastal village. Expectations immediately arose that the wreckage was from the lost PB4Y-2 patrol plane. Unfortunately, a check of known crash sites identified the wreckage as a C-47 cargo plane that had crashed during World War II.

To date no evidence of the lost plane or its twelve-man crew has ever been found.

Chapter Ten

November 4, 1948

Fading Signal

Light rain peppering the windshield ran in rivulets off the sides from the strong air stream. A few drops pooled around the edge and seeped through the rubber seal into the DC-3's cockpit, while gusting thirty mph winds buffeted the aircraft with repeated jolts of turbulence. The two pilots barely noticed as they listened intently through their headsets for the dots and dashes that identified the airway signal. Time dragged by in slow ticks of the dashboard clock as they concentrated on tracking the faint reception above the cold waters of the Alaska coast. Outside the dim glow of the instrument lights the night was dark and foreboding, as if warning of danger ahead. Higher clouds blocked any celestial illumination from reaching the aircraft and a layer of lower clouds effectively masked all visual contact with the ocean below.

In spite of the turbulence most of the fifteen passengers in the cabin were trying to sleep, either pretending to ignore the rough air or fighting the onslaught of airsickness. The remaining few attempted talking over the loud drone of the twin Wasp radial engines or stared outside their small windows into the black sky. Seating was cramped but as comfortable as could be expected for the limited space available in the forward cabin. The rear seats had been removed and the area instead configured for carrying cargo and luggage. There was no stewardess on board, only the two pilots flying the plane who alternated checking on the passengers every thirty minutes or so, offering a limited choice of coffee and stale donuts.

Proceeding outbound on the southeast leg of the Yakutat Radio Range for almost an hour, pilots Andrew Kinnear and Richard Wilson monitored the two distinct Morse code signals and station identifier being broadcast simultaneously over the radio frequency. At first the separate audio tones were steady and strong; mixing together in a repetitious whine that signified the aircraft was tracking the center of the airway. As the distance from the range station increased, the tone began fading and the static became more intense, until the signal was no longer recognizable. Continuing the flight by dead reckoning over what they hoped was still the airway course, the pilots began cross-tuning to the frequency for the range station at Gustavus, trying to identify the southwest leg where it intersected the airway between Yakutat and Sitka at Cape Spencer Intersection. Once the intersection was fixed, the Sitka Radio Range frequency would be selected to continue tracking the airway inbound on the northwest leg of the Sitka Radio Range.

Alaska Airlines DC-3 over the unforgiving coastal mountains. (Museum of Flight-Seattle, WA)

Neither pilot was overly concerned about the momentary loss of reception, as it was not an uncommon occurrence, especially when flying in poor weather conditions. Their confidence was rewarded several minutes later when the Gustavus radio signal was faintly heard through the headsets, barely recognizable over the constant clutter of background static, but enough to verify they were passing through the southwest leg of the station. A routine position report was transmitted to Gustavus Radio at 5:10 am Alaska Standard Time, confirming they were over the intersection at 10,000 feet and estimating an arrival over Sitka at 5:44. One hour and three minutes had passed since departing Yakutat. Nothing was heard from the flight again as it mysteriously vanished somewhere along the stretch of remote coastline.

The aircraft was a DC-3C belonging to Pacific Alaska Air Express on a flight from Anchorage, Alaska to Seattle, Washington. It initially departed Anchorage the previous night at 10:26 pm with ten passengers. Stopping at Homer on the south end of the Kenai Peninsula, the plane picked up four additional passengers before continuing southeast along the coast, landing in Yakutat at 2:57 am.

Chapter 10—Fading Signal

While on the ground the plane was serviced and one last passenger added before it lifted off the runway at 4:07 am with full fuel tanks. Flight time en route to the next scheduled stop at Annette Island should have been approximately three hours, based on forecast winds.

Routine position reports were required passing the Cape Spencer Intersection and Sitka Radio Range. When an hour and forty-five minutes had passed since the flight's last transmission over the intersection and an hour after its estimating arrival over Sitka, an emergency warning order was issued by the Civil Aero-

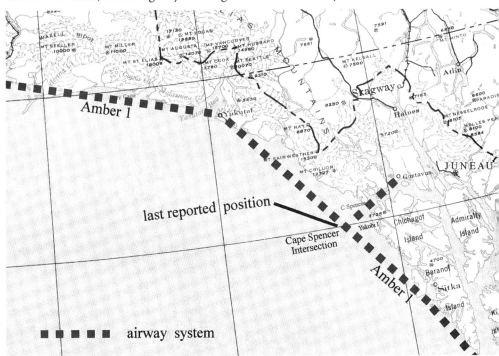

Southeast Alaska map showing the coastal airway route used by the missing DC-3.

nautics Administration.[1] All stations between Seattle and Anchorage were notified, directing them to attempt contact with the missing plane. With no response forthcoming, it was first presumed the Pacific Alaska Air Express DC-3 experienced some sort of radio difficulties that prevented communication. When it also failed to arrive at Annette Island, search and rescue units and officials at the Civil Aeronautics Board were notified.[2]

---

[1] The Civil Aeronautics Administration was the predecessor to the current Federal Aviation Administration.
[2] The Civil Aeronautics Board was the investigative branch of the Civil Aeronautics Administration and the predecessor to the current National Transportation Safety Board.

The 10th Rescue Squadron in Anchorage and Coast Guard facilities at Sitka and Juneau were the first to respond. A Coast Guard amphibian aircraft was the first to reach the area around Cape Spencer Intersection by mid-morning and a second Coast Guard amphibian arrived a short time later. They were joined in the search by an Air Force C-47 from the 10th Rescue Squadron that afternoon, but no evidence of the plane or its occupants was found.

Weather conditions around the area of the missing DC-3's last known position consisted of variable cloud layers down to 2,000 feet, with light rain showers that reduced visibility to only a few miles. Often flying just below the cloud base, the

Wien Alaska Airlines DC-3 at a remote location in Alaska. (Randy Acord)

search and rescue aircraft could only cover one narrow corridor above the heavy seas at a time, maintaining a basic compass course on a back and forth, parallel search pattern. The larger the distance flown over water, the larger the gaps in coverage that occur due to heading and turn errors. Combined with high, rolling waves that could easily hide large objects at more than a few hundred yards, the chance of spotting scattered pieces of wreckage or bodies required as much luck as skill. For the search crews to have any realistic chance of success they needed to be in close proximity to what they were looking at, and for the object to draw their attention from the surrounding colors and shades of the violent wind-tossed seas.

By the second day of the search a Coast Guard PBY from Washington and a SB-17 from California configured for search and rescue missions were dispatched to join the limited operation, aided by a few commercial aircraft from Sitka and Juneau. Unfortunately, most of the military aircraft and ships that would normally have been available were diverted to two other ongoing search operations in the Aleutians and along the coast of British Columbia.

The air route between Yakutat and Sitka flown by the missing DC-3 was divided into sectors and assigned to the various search aircraft involved for maximum coverage of both land and sea. A wide swath on either side of the airway was covered as effectively as possible, considering the weather and factoring in currents and winds that would have dispersed evidence of the plane over miles of ocean and along the coastal islands. On a lesser degree the air route from Sitka to Annette Island was also searched, in case the missing aircraft had continued past Sitka without transmitting a position report.

If the Pacific Alaska Air Express DC-3 had lost power and was forced to ditch

SB-17 search and rescue aircraft. Note the air droppable boat mounted underneath the fuselage. (Dan Lange)

in the Gulf of Alaska, survival of the occupants was dependent on a successful evacuation into the plane's life rafts and minimal exposure to the numbing water and harsh winds prevalent in the area. Even then, without heat, shelter and proper cold weather clothing, survival for more than a few hours was unlikely. But if the aircraft had gone down on one of the islands or the occupants reached shore safely, the chance of prolonged survival and eventual rescue was dramatically increased. In addition to the life rafts, the DC-3 carried a two week supply of emergency rations, fishing gear, a rifle, axes, blankets and two portable emergency radios.

Search operations continued through 24 November with Coast Guard and Air Force 10th Rescue Squadron aircraft able to cover several thousand square miles of rough seas, forested islands and coastal mountains. Weather fluctuated between clear flying conditions on a few days to low clouds and heavy snow or rain showers during most of the days. On several occasions planes could not fly at all due to foul weather hanging over the search area. Not one piece of wreckage from the missing DC-3 was ever found.

Small debris from an unknown type of aircraft was located during the search on Chicagof Island, 184 miles southeast of Yakutat, but the length of time the pieces had been there could not be determined and also could not be identified as positively coming from the Pacific Alaska Air Express flight. There was also a report of a person on Prince of Wales Island, much further south of Sitka, hearing an aircraft crash in the hills on the day the flight disappeared. The report was investigated and was not believed to be credible.

Alaska Airlines DC-3 sits ready, unfazed by the winter weather. (NOAA)

Months after the search operation was officially terminated military and commercial aircraft flying the Yakutat to Annette Island route kept a sharp lookout for the lost DC-3. Eventually its fate was forgotten along with the seventeen men and women aboard, as newer and more dramatic disasters continued to capture the headlines. The disappearance was not the first by a large aircraft off the Southeast Alaska coast and unfortunately it would not be the last.

The investigation into the accident by the Civil Aeronautics Board (CAB) could not determine a likely cause of the disappearance. Even though rain, high winds and turbulence were prevalent along the route, they were not considered significant enough to affect the flight. Icing was also not believed to be a factor, as conditions at the cruising altitude of 10,000 feet were not conducive for either carburetor or structural icing. A full weather briefing was received by the pilots in Anchorage prior to departure, detailing the conditions they would have encountered during the flight.

Both Captain Andrew Kinnear and co-pilot Richard Wilson were experienced pilots who had flown the same route together on four previous occasions. They each had three days of rest before the flight.

Kinnear served with the Army Air Forces in the China-India Theatre during World War Two and had been flying in Alaska for two years. He held an Airline Transport Pilot certification and possessed 3,600 hours of total flight time, with 3,200 of those in multi-engine aircraft.

Company records indicated Wilson held a valid Airframe and Engine license and a pilot's multi-engine and instrument rating. His total time was 1,800 hours, with 325 hours in multi-engine aircraft. The Civil Aeronautics Administration (CAA), however, only showed Wilson certified as a student pilot. Whether it was a paper mix-up, typing error or fraud was unknown, but his A&E license number on file at the CAA was the same number shown on a copy of an airmen

Northwest Airlines DC-3 in fair weather far from Alaska. (Museum of Flight, Seattle, WA)

certificate at Pacific Alaska Air Express. A&E or Airframe and Engine licenses have nothing to do with a pilot's qualification. They are only issued to mechanics certified to work on aircraft.

Maintenance records for the aircraft could not be examined by the CAB due to the fact they were aboard the DC-3 when it disappeared. The previous crew who flew the aircraft on 2 November stated the plane and engines were in satisfactory condition with no mechanical deficiencies, and no reports were received from Kinnear or Wilson indicating any problems.

The Pacific Alaska Air Express DC-3C had 4,320 total hours accrued on the airframe since being manufactured in June 1943 and 348 hours since its last major overhaul. The engines were Pratt and Whitney R-1800-S1C3G Twin Wasp radials, providing a normal cruise speed of 207 mph and range of 2,125 miles.

Originally designated as a C-47 when it first entered service with the U.S. Army Air Forces, the aircraft was later sold as surplus following the end of World War Two. All C-47s converted for civilian use were then given a new designation of

DC-3C. The airframes remained basically the same after conversion except for minor structural modifications to the door, windows and seats for passenger comfort.

In the two days following the disappearance of the Pacific Alaska Air Express DC-3, the CAA conducted flight checks of the radio ranges at Sitka, Gustavus and Yakutat. All three were found to be operating within normal tolerances. By then, however, the weather conditions had changed enough where any signal irregularities would not have been duplicated.

The CAB determined the missing plane was in satisfactory operating condition upon departure from Yakutat and the flight was routine up until its last position report. When the aircraft made its last report over Cape Spencer Intersec-

Alaska Airlines DC-3 over the coast of Southeast Alaska. (Museum of Flight, Seattle, WA)

tion, it still had eight hours of fuel remaining onboard. The probable cause of the accident could not be determined.

Several possibilities existed that could have caused the accident, but they are speculative and not based on any concrete evidence. One has to do with the low-frequency radio range systems used for instrument navigation at the time of the disappearance. A navigational error by the pilots is another factor, as is the occurrence of unexpected and extreme weather conditions being encountered between Yakutat and Sitka. Each possibility could have contributed to the accident individually or in combination with each other.

In general, low frequency radio ranges were noted for their individual idiosyncrasies that could include multiple signals, bending course legs and unpredictable changes in signal strength. Those problems often intensified during periods of precipitation static when flying in rain or snow.

Navigation along a radio range leg or airway was accomplished by monitoring audio signals transmitted by a ground station. Sectors on each side of the airway were identified by distinct audio tones, transmitted in a series of either "A" or "N" Morse code dots and dashes. Each was broadcast simultaneously in con-

Chapter 10—Fading Signal

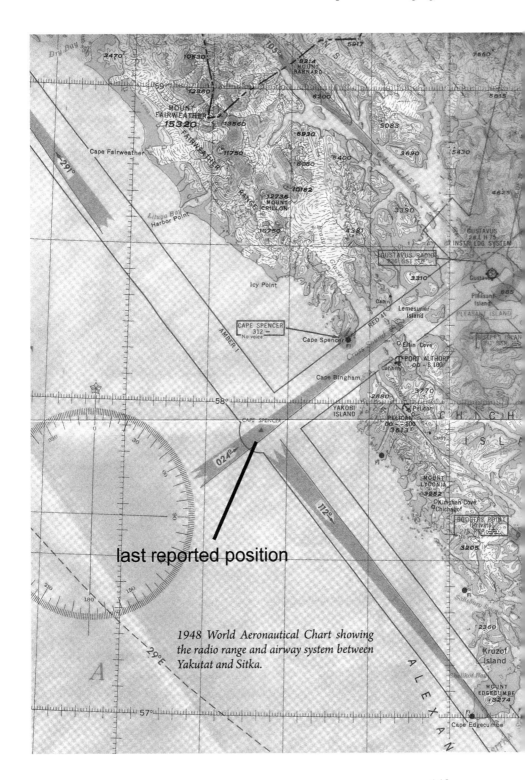

last reported position

1948 World Aeronautical Chart showing the radio range and airway system between Yakutat and Sitka.

junction with the particular station identifier. In perfect conditions, when an aircraft strayed to one side or other of the airway, the "A" or "N" tone for that sector became more predominate, warning the pilot he was flying off course. Tracking the correct on-course signal was thus accomplished by referencing the dots and dashes filtering through the static of the low frequency receiver. As an aircraft approached a range station the signals would gradually intensify and when flying away from the station the signals would slowly fade. In poor weather conditions, however, adequate reception was often unreliable or completely unreadable, giving inaccurate information to determine the aircraft's position and course track.

If the pilots of the lost Pacific Alaska Air Express DC-3 experienced reception irregularities, they could have been unaware they were flying off course into higher coastal mountains north of the airway. A military C-54 in 1950 and a civilian DC-3 in 1952 crashed in the mountains between Yakutat and Spencer Intersection after straying off the airway during foul weather. Another civilian airliner disappeared in the same area in 1951 under similar circumstances and several smaller planes have vanished over the ensuing decades. Two military aircraft, a Navy R4D-1 in 1942 and a Navy PB4Y-2 in 1945, both on routine passenger flights well away from any combat zone, were also lost along the coast and have never been found.

For the Pacific Alaska DC-3 to have flown off the airway into the higher coastal mountains would have required a minimum navigation error of thirteen degrees. The error would have taken the plane approximately fifty miles north of Cape Spencer Intersection into Mount La Perouse, the closest mountain above its flight altitude. Increased degrees of error would have caused the aircraft to fly further inland into any of several higher peaks within the Fairweather or St. Elias Mountain Ranges.

Since the crew did in fact report over Cape Spencer Intersection, and assuming their position was accurate, they would have already been in a location south of the higher coastal mountains. From that point even a navigational error of fifteen degrees would have kept them on a southeast track over the various islands of the archipelago for hundreds of miles, none of which had mountains higher than 10,000 feet. Dozens of lesser mountains reached 4,000 feet in elevation, but only two were above 5,000 and none extended above 5,400 feet.

Winds at the plane's flight altitude were forecast at 30 knots from 190 degrees between Yakutat and Sitka. Based on the winds and a cruise speed of 207 mph, the ground speed would have been reduced to 200 mph with a required wind correction of 10 degrees right. En route time from Yakutat to Cape Spencer Intersection should have been 54 minutes. Actual flight time was 63 minutes, indicating stronger than predicted in-flight winds or a shift in the winds more in line to the aircraft heading.

Chapter 10—Fading Signal

A change in the airway course at Cape Spencer Intersection might also be of significance. The intersection was the dividing point between the Yakutat and Sitka Radio Ranges and was also the point where the airway altered course by two degrees, allowing a direct bearing inbound on the northwest leg of the Sitka Radio Range. If the pilots failed to make the two degree change in heading, the error would not by itself been significant, but if combined with inaccurate signal reception, misinterpretation of their actual position, stronger than predicted winds from the west and insufficient wind correction, the difference could have been disastrous.

Light to moderate turbulence was predicted over the route between Yakutat and Sitka at 10,000 feet, which was of little consequence except the comfort of the passengers. DC-3 aircraft were known for their durability and could easily handle rougher weather conditions. It is possible, however, that the aircraft could have encountered violent wind sheers that caused a sudden and catastrophic structural failure. Unpredictable mountain waves were not frequent, but were common enough in Alaska that they were referred to by their own name. "Williwaws" have claimed many unsuspecting aircraft over the years in Alaska. The southeast coast is no exception.

During the 1950s, military aircraft flying across the Gulf of Alaska, much further offshore than the commercial airway, reported experiencing powerful and unpredictable westerly winds as high as 120 mph.[3] The sudden and dangerous winds occurred without notice, pushing aircraft as much as a hundred miles off course before the crews correctly interpreted conflicting navigation signals. If an aircraft flying the coastal commercial airway, only twenty miles west of the high mountain ranges, encountered similar wind conditions, it could easily have been pushed inland before the crew was aware of what was happening.

If the pieces of wreckage found on Chicagof Island were from the missing DC-3, it would suggest a powerful external force or some internal structural failure destroyed the aircraft without warning. No problems with the aircraft, navigation or weather were reported by the crew, and no distress calls sent before it disappeared. Radio communication appeared normal at the time of the last position report.

What really happened to the Pacific Alaska Air Express will probably never be known. The thousands of square miles of towering snow-capped mountains, immense glaciers, thick forests and deep waters of Southeast Alaska can hide many secrets. It's still out there waiting to be found, perhaps sitting atop a high peak, far inland of the coastal range, or on one of the heavily timbered islands, just around the corner of nowhere. Only time will tell.

---

3 The phenomenon is a result of diverging winds from the high altitude jet stream, which have been documented to occur unexpectedly as low as 8,000 feet.

Chapter Eleven

January 26, 1950

Without a Trace

A long procession of soldiers and airmen began filtering off the two military buses as they drove alongside and stopped near the large transport's starboard wing on Elmendorf Air Force Base in Alaska. Two military dependents and a civilian technician were the only non-military personnel among the group of thirty-six passengers and eight crewmen. Both dependents, one the young pregnant wife of an Air Force Master Sergeant and the other her infant son, were returning to the States for the birth of her second child, while the remaining passengers were being reassigned to new duty stations or taking a few weeks of well deserved leave. All of them appeared exhilarated by the prospect of seeing loved ones and friends again, but no one was looking forward to the many long hours of cramped conditions aboard the plane and bouts of rough air during the long flight still ahead.

The aircraft assigned for the flight was a Douglas C-54D Skymaster belonging to the 2nd Strategic Support Group, 97th Bomb Wing of the 8th Air Force, under the jurisdiction of the Strategic Air Command (SAC) at Biggs Air Force Base near El Paso, Texas. It was a military version of the well known Douglas DC-4, originally designed for commercial passenger service in the civilian airline industry. The military version incorporated several modifications for long-range airlift capability and heavier capacity weight demands required for large internal loads. A larger cargo door, additional structural supports, removable hoist and increased fuel capacity were the most significant additions. Both military and civilian versions had a pressurized cabin, which at the time of its design was a relatively new commodity.

Commanding the plane was First Lieutenant Kyle McMichael, an instructor pilot who would be evaluating pilots First Lieutenant Mike Tisik and Major Gerald Brittain. The rest of the eight-man crew consisted of navigator First Lieutenant Joseph Metzler, radio operator Staff Sergeant Clarence Gibson, and crew chiefs Master Sergeant Clyde Streitmahn, Technical Sergeant Harry McConegley and Staff Sergeant Raymond Snow.

C-54 aircraft first entered service with the U.S. military in 1942 during the early years of WWII. More than 1,200 C-54s of various configurations were eventually placed in military service. The D model had a maximum ceiling of slightly more than 22,000 feet, which was attained without the requirement of supplemental oxygen thanks to a pressurized cabin. A range of 3,100 miles was possible with the C-54's normal fuel capacity. Some C-54 aircraft were still serv-

Alaska Air Command C-54D preparing for departure at Elmendorf AFB, March 1954. (3rd Wing Historical Office, Elmendorf AFB)

ing in regular Air Force units until 1962, while others stayed in service with Air National Guard units through the 1960s, before eventually being replaced by more modern transports.

As the thirty-six passengers were seated in the cabin and briefed by the crew, the pilots signaled through the cockpit window to the ground personnel outside they were ready to start the engines. Each of the four Pratt and Whitney R-2000-11 radial piston engines was brought to life in a belch of smoke and slow turning propellers as an airman stood nearby with a fire extinguisher. The black exhaust quickly changed to a dull gray cloud as the powerful 1450 hp engines spun the three-bladed propellers into a whirling circle of transparent motion.

Once each engine was running smoothly, verified by cockpit indications of temperature, pressure and rpm, the pilots signaled the ground personnel to remove the wheel chocks so they could get underway. The four throttle levers mounted on the center console were pushed forward in unison, increasing power to the engines as the aircraft moved along the taxiway and into position for departure.

Chapter 11—Without a Trace

USAF C-54D being checked by ground personnel prior to flight. (Museum of Flight, Seattle, WA)

An instrument flight clearance from Elmendorf AFB to Great Falls, Montana was received by the crew as they maneuvered toward the runway. The route would take them north from Anchorage until intercepting the Green 8 airway, then northeast at 11,000 feet to Gulkana and Northway near the Alaskan/Canadian border. Over Northway they would intercept the Amber 2 airway and proceed southeast at 10,000 feet, crossing the Canadian cities of Whitehorse, Fort St. John and Calgary before their arrival in Great Falls, Montana, some eight and a half hours later.

USAF C-54 over an inland mountain range. (Museum of Flight, Seattle, WA)

It took several minutes for the C-54 transport to reach the end of the runway. Once properly aligned and ready for departure, additional cockpit checks were completed and a final takeoff clearance received from the tower. As the engines were brought to full power the pilot relaxed his pressure on the brake pedals, allowing the aircraft to accelerate forward down the runway. Its speed increased slowly at first, then became faster as the aircraft gained momentum and rose in a steep, climbing turn across the waters of Cook Inlet into the cold winter sky.

The time was 11:16 am Alaska Standard Time (AST) as the C-54 headed north from Anchorage for the long flight through Canada to the United States. Weather conditions for the route were not expected as much more than a minor inconvenience. The forecast called for mostly clear skies with a few scattered clouds along the 2,000 mile course, with the exception of overcast conditions at 7,500 feet and possible icing in clouds near the city of Whitehorse in the Yukon Territory. No strong turbulence or icing was predicted over the remainder of the route.

As the aircraft flew east toward Gulkana, paralleling the rugged, snow-covered peaks of the Chugach Mountains to the south and Talkeetna Mountains to the north, the crew and passengers must have been captivated by the seemingly endless expanse of towering ice-covered pinnacles that surrounded the deep glaciers and snow-filled valleys. In every direction high jagged peaks could be seen jutting from formidable mountain ranges that encompassed the horizon, many rising to heights several thousand feet above their flight altitude. It was a spectacle few people outside of Alaska would ever experience, but one that once witnessed would forever be remembered.

Thoughts of home were momentarily forgotten as the faces of the passengers stared through the windows in wonderment at the breathtaking scenery; both beautiful and threatening at the same time. As spectacular as the view was from the plane, many were still thankful to be leaving the rugged, isolated territory of Alaska behind. Enjoying a flight over pristine wilderness was one thing, living in a land of often extreme temperatures, few luxuries, high prices and dangerous wildlife was more than most people were willing to tolerate for very long. It took a special breed of men and women to not only live in Alaska, but to appreciate and accept the region's independent spirit as their own.

After passing over the Gulkana range station and turning northeast to follow the airway, the flight crossed the eastern stretch of the Alaska Range only minutes before reaching Northway on the Tanana River, twenty-nine miles from the Canadian border. The signal reception from the range station grew stronger with each approaching mile, then silent as the aircraft passed directly over the "cone of silence".[1] As the aircraft turned southeast on the outbound leg of the range and intercepted the Amber 2 airway, the signal was heard again, only this time slowly fading in intensity as the distance increased. It crossed the Canadian border minutes later a few miles north of the Alaska Highway and continued on course. At 1:09 pm AST, almost two hours after departing Elmendorf AFB and seventeen minutes after passing Northway, the flight called crossing the Snag Radio Range station, located near Beaver Creek, Yukon Territory.

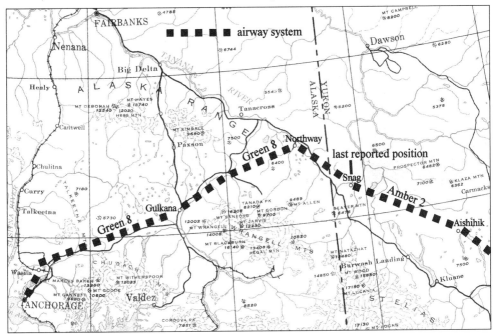

Airway route from Anchorage, Alaska to Aishihik, Yukon Territory.

The radio transmission received by the station's radio operator stated an estimated time en route to the next reporting point at Aishihik of twenty-eight minutes. There was no indication of a problem with the aircraft or weather. The Aishihik Radio Range was ninety-two nautical miles east of Snag and forty-five miles northeast of Burwash Landing on the Alaska Highway. No further radio transmissions were received from the plane after passing Snag. Attempted contact by various communication stations along the C-54's route of flight over the next seven hours were without success. When the plane failed to arrive at its

---

1 Radio range stations were navigational aides used for instrument flight from the late 1930s through the 1960s. Each radio range was comprised of four towers transmitting the station identifier and course signal on a separate leg or beam that could be monitored by a radio receiver in the aircraft. When passing directly overhead the station's inner tower zone or "cone of silence" between the four towers, the signal was lost until the aircraft continued back outside the zone.

destination in Great Falls, Montana, the plane was officially reported as missing. No evidence of the aircraft, its crew or its passengers has ever been found.

After the aircraft was reported missing an intensive aerial and ground search was immediately organized by American and Canadian military forces in the region. Overall control of the search operation was placed under the joint jurisdiction of the United States and Royal Canadian Air Forces. Numerous military

USAF C-54s at Kotzebue, Alaska, 1959. (Randy Acord)

aircraft and personnel were already deployed at Whitehorse, Yukon Territory as part of a major American and Canadian winter training exercise and were made available for the search. Additional aircraft were dispatched from bases in Alaska, Canada and the continental United States.

The equipment and personnel already in place for the joint training exercise in Whitehorse allowed for a quick and easy transition to search for the missing plane and provided an immediate communication network that would not have otherwise been available. Transportable navigation aids and weather facilities used for tactical deployments were also on hand for the training exercise and activated for the search operation. A major portion of individuals involved during the critical first few weeks were the soldiers and airmen participating in the winter training exercise.

Both Whitehorse and Fort Nelson, British Columbia were used as staging bases for search aircraft and ground teams, with most of the assets eventually being moved to Whitehorse where they were placed under the direct control of a joint operation headquarters staff. U.S. Air Force units provided the majority of aircraft and personnel involved in the search, with U.S. Army, Canadian military and civilian components assisting as needed.

A RCAF plane was the first to begin searching for the missing C-54 on the night of 26 January. It concentrated on an area north of Edmonton, Alberta

C-54 of the USAF 21st Composite Wing over a winter landscape in Alaska. (3rd Wing Historical Office, Elmendorf AFB)

where a false report placed the missing aircraft earlier in the day. The same night three USAF B-17s, five C-47s and a C-54 were dispatched to Whitehorse and Fort Nelson to begin searching the next day.

Reports of flares being sighted on the Alaska Highway near Watson Lake were received on 27 January, giving false hope for a quick rescue of the missing plane's occupants, but the flares were soon identified as coming from a stranded vehicle. Another erroneous report claimed the lost C-54 radioed it was passing Whitehorse before disappearing, causing the search to initially focus between Whitehorse and Fort Nelson, instead of nearer the Alaskan border. Severe weather conditions from a strong frontal system moving over the Yukon Territory hampered initial search efforts, but over a dozen Canadian and USAF planes were able to get airborne and cover more than 7,500 square miles the first day. Communication stations in western Canada were also advised to listen for distress calls in case survivors were able to transmit their location on one of the survival radios, and inhabitants in the area were asked to be on the look out for possible flares or smoke signals.

When no new leads were forthcoming after two days the consensus was the missing C-54D was most likely down somewhere between Snag and Whitehorse, with a lesser possibility of it being down in the Watson Lake area. Those two locations became the primary focus of the search. More aircraft, including amphibious planes and helicopters arrived in Canada the same day, doubling the

Air Transport Command C-54 parked at Boeing Field in Seattle. (Museum of Flight, Seattle, WA)

amount of aircraft already available. Dozens more were still en route from bases as far as the southern United States.

The biggest priority of the operation was in locating survivors of the missing plane, especially after being exposed to conditions of extreme cold temperature and high winds for a prolonged period. Even though the plane carried emergency food supplies for several days and Arctic clothing for all the occupants, survival in the harsh environment would also depend on the availability of adequate shelter.

More than fifty aircraft were involved in the search by 29 January and were able to cover a large area from Snag southward. Other aircraft en route from Great Falls, Montana searched a wide stretch northward.

Even as more and more aircraft arrived in Canada daily, a vast portion of the search area could not be observed effectively from the air. Thousands of square miles of uninhabited land were covered with thick timber, while other areas were filled with remote mountain valleys and ice-layered peaks that could easily hide a hundred different aircraft. Unless the missing plane remained relatively intact and in an open area that was clearly visible from the air, the chance of sighting the wreckage was extremely limited. From a search altitude of several thousand feet an object the size of a four-engine transport would only appear as a tiny speck on the landscape. A complete silhouette of an aircraft on the ground would at least give the search crews something to key on, but more often than not they were simply looking for a glint of metal, patch of broken trees, recent snow slide or discoloration that could have been caused by impact or fire. With thousands of square miles to cover, search crews only hoped they could pass close enough to the crash site for survivors to send some sort of signal.

The entire search area was divided into multiple thirty square mile grids which were assigned to one or more aircraft. Each aircraft would then fly a parallel pattern back and forth across the grid while maintaining a two to five mile separation, depending on the type of terrain. Additional crew members were carried aboard each aircraft as observers, so as many sets of eyes as possible were searching the terrain in overlapping fields of view.

With numerous aircraft involved in the search and many flying in marginal weather conditions for extended periods of time, the endurance of both planes and crews was stretched to the limit. It was inevitable more accidents would occur. Four days into the search a C-47 crashed on a mountain near Carcross, southeast of Whitehorse, when it was caught in a strong downdraft. Luckily no one was fatally injured and the six-man crew was rescued the next day. Several other aircraft were damaged during landing attempts by fatigued pilots who had been flying with little sleep for days.

On 30 January, clear weather conditions prevailed over much of the search area for the first time since the C-54D disappeared, allowing dozens of aircraft to cover more than 40,000 square miles of wilderness. By the end of the day almost fifty percent of the designated grids had been searched at least once. The remaining grids consisted of high mountain ranges still masked by heavy cloud cover, including dozens of towering peaks above 7,000 feet within fifty miles on either side of the missing transport's flight path. All were surrounded by hundreds of lesser, but no less formidable mountains and hundreds more heavily forested hills and glacial filled lakes.

The same day a Forest Ranger came forward to report seeing a large, low flying aircraft during the afternoon of 26 January, only forty miles southwest of Snag. He said the aircraft disappeared shortly before he heard a loud explosion in the same direction, followed by columns of dark smoke. Another person in the area claimed to have seen a two or four-engine aircraft flying along the highway the same day in a southeast direction, while a further report hundreds of miles away in British Columbia claimed several people saw an aircraft fly overhead shortly before hearing a loud explosion. All those areas were extensively searched, but no evidence of the plane was found.

Numerous sightings of smoke signals in the mountains along the Alaska/Canada highway were reported over the next several weeks, usually coinciding with periods of high wind activity. All the sightings were investigated and were either unsubstantiated or determined to be blowing snow off exposed mountains and ridges.

Other reports on the whereabouts of the missing plane were almost comical if not for the seriousness of the situation. Telegrams and letters were received from psychics and Ham radio operators thousands of miles away, claiming various locations throughout Alaska and Canada where the missing plane could be found. One Ham radio operator, who supposedly monitored a distress call from the survivors, placed the missing transport in the Azores near the African coast. Another radio operator provided coordinates that were located in Denmark. In each case every legitimate report was taken seriously and investigated.

Temperatures plummeted well below zero across western Canada and the northwest United States during the last few days of January, causing extensive maintenance delays among participating aircraft. Several planes were stranded at smaller airfields and others were moved to Edmonton and Calgary to utilize more adequate repair facilities and hangar space. The search still continued even under the most extreme conditions, and by the first of February more than ninety percent of the grids between Snag and Fort Nelson had been covered at least once, while more than fifty percent of the grids were covered twice and approximately a third of the grids three times. With each additional flight the chances of finding the plane or survivors improved dramatically.

Weak and unreadable radio transmissions were received over the emergency band frequency by communication stations and aircraft during late January and early February, prompting a radio sweep of the entire route from Great Falls to the Alaskan border by ten specially equipped electronic surveillance Air Force B-29s. It was believed survivors from the missing plane might have been using a hand-cranked "Gibson Girl" emergency survival radio to transmit distress calls. Designed as an emergency line-of-sight device during World War Two for use in desert or ocean areas, the portable radio had a limited range that was easily distorted by higher terrain. An official involved in the search was quoted by a local newspaper as saying an aircraft would need to fly almost directly over a "Gibson Girl" signal in mountainous terrain for any transmission to be heard.

None of the strange signals monitored on the emergency frequencies were identified as coming from the lost aircraft during the B-29s' electronic sweep, but several transmissions were pinpointed near Seattle, Washington and determined to be unrelated.

Chapter 11—Without a Trace

On 7 February a stronger series of emergency signals were monitored by numerous aircraft and communication stations throughout Alaska and Canada in an area approximately eighty-five miles northwest of Whitehorse. It was a promising development that turned into a separate emergency when the signals were identified as coming from another crashed C-47 aircraft involved in the search. Similar to the accident on 30 January, the plane had encountered strong downdrafts over the crest of a high mountain and was unable to regain altitude before impacting the terrain. All ten occupants survived and were rescued a few days later by military helicopters.

Weather conditions over the search area varied from clear skies to a thick overcast during the first two weeks of February, often accompanied by heavy snowfall and below zero temperatures. Aircraft continued flying on a daily basis and only remained on the ground when the ceiling and visibility would not allow operations from the airfield.

Additional leads continued to develop through February, but none could be substantiated after extensive visual searches of the areas. Three separate witnesses came forward in February, all from different locations, who reported hearing a loud explosion in the vicinity of Caribou Mountain near Carcross, Yukon Territory on 26 January. Their account of the crash was believed to be a simple confusion with the C-47 that crashed on the same mountain on 30 January. Even so, several aircraft were dispatched to search the area again, with no results.

Another potential lead came from a trapper near Burwash Landing, who claimed he heard a large airplane go down near his camp on 26 January, and who observed a recent landslide on the mountain the next day. Four aircraft and a ground search team covered the area extensively, but could find no evidence of wreckage.

Military direction finding stations in Canada continued monitoring occasional distress signals over the emergency band frequency for weeks, but they could never be identified. By the middle of February it became apparent the missing Air Force C-54 would probably not be found until at least the next summer when melting snow could reveal its location. Hope of finding any survivors had already been dismissed as an unlikely possibility.

On 14 February a new search for a missing USAF B-36B bomber near the coast of British Columbia, diverted many of the aircraft involved in the C-54 search away from Whitehorse. Only a few aircraft remained in the area over the next several days and their flights were significantly reduced. One of those planes crashed on 16 February after losing power during takeoff from a frozen lake near Snag. No one was seriously hurt when it went down in the trees, but it was the third C-47 to crash since the operation began.

By 17 February the remaining aircraft involved in the search for the missing C-54 flew only if a new lead materialized. By 20 February the entire operation was terminated. All military aircraft in the region were reassigned to the B-36 search or returned to their home bases. It marked the end of the most intensive aerial search ever conducted in Canada and the United States at the time. More than 165,000 square miles of territory were searched multiple times between 26 January and 20 February. As many as fifty different aircraft and hundreds of military and civilian personnel were involved on any given day. Nothing was ever found.

There was no information in the official Accident Report which could explain the disappearance. The missing C-54's maintenance records did not indicate a prior problem with the airframe or its engines, problems which might have caused it to crash unexpectedly. Only five years old, the plane was relatively new

The missing C-54D Skymaster, #42-72469, over Alaska on an earlier flight. (Don Downey via Air Classics magazine)

and had flown roughly 4,400 hours since leaving the factory. Three hundred and fifty of those hours were after its last overhaul in August the previous year. The pilot-in-command was an experienced aviator with an instructor rating, having flown almost 2,500 flight hours in his career, including more than 1,300 hours in the same type aircraft. Two other pilots, a navigator, radio operator and three crew chiefs made up the rest of the crew. All were equally experienced.

Aside for the overall experience of the crew in flying Air Force transport missions, none were experienced in winter or Arctic flying conditions that frequented Alaska and the Yukon. This relevant piece of information was not discussed in the Accident Report and could have been a contributing factor in the aircraft's demise. Stronger than normal weather systems, large variations between magnetic and true north, and limited navigational and communication coverage are problems not easily overcome in the northern regions. Every radio range station across North America had its own idiosyncrasies, from false signals to mysterious bends in the course leg, to a complete loss of signal for indefinite periods, all made even worse by poor weather and static interference. Pilots often learned from each other the particulars of each radio range station, but advice alone was not always enough. First encounters, especially in bad weather, were best experienced with a knowledgeable mentor in the opposite seat.

One theory proposed by a retired Air Force officer in 1994 suggested the weather conditions present during the flight could have caused the C-54's crew to become disoriented and stray off course without realizing it. He believed 50 knot northeast winds which existed at the plane's flight altitude near Snag, could have pushed it well south of the airway into the St. Elias Mountain Range, somewhere below Kluane Lake in the Yukon Territory. Higher level mountains in that area reach well above 10,000 feet.

With a direct 50 knot crosswind from the north and no correction applied, the aircraft would indeed have been pushed 17 degrees right of course along a track south of Kluane Lake, into the higher mountains of the coastal range. The crew, however, was aware of and should have been compensating for the strong in-flight winds, since their actual arrival time over Snag and estimated time to Aishihik were faster than would normally be flown by as much as twenty-three knots.

Even though the aircraft could still have been pushed off course to a lesser degree without the crew noticing the error, especially with low clouds blocking any visual sighting of the ground, other factors would also have had to of been in play. In most fatal accidents the cause is usually not from one factor alone but several factors acting together. If a wind correction error was combined with one or more radio range errors, causing the crew to misidentify their position, the actual track could easily have been more than 17 degrees south of the intended course. An overcompensation error in the opposite direction could have sent the transport hundreds of miles north, into a vast expanse of uninhabited lakes and tundra.

Since the plane did send position reports over several locations on the airway system between Elmendorf AFB and Northway, it is doubtful if an error would have gone unnoticed. The sky conditions were relatively clear at the time, providing good visual references with the ground. Any error would therefore seem to have occurred sometime after passing Northway, when heavier accumulations of clouds would have been encountered.

The evidence suggests the accident was both sudden and catastrophic in nature. Whether caused by structural failure, navigational error or crew incapacitation, the result was the same. No attempts were made to obtain a direction bearing or fix on their position and no distress calls were sent at any time during the flight. The large four-engine transport simply vanished without a trace.

Then again, what if some or even one of the dismissed sightings was accurate? Could it be the aircraft crashed or ditched with inadequate time to send a distress call, only to have survivors vainly attempt contact over the next several weeks, all the while slowly succumbing to the harsh elements and diminishing food supplies? One can only imagine their frustration at hearing search aircraft

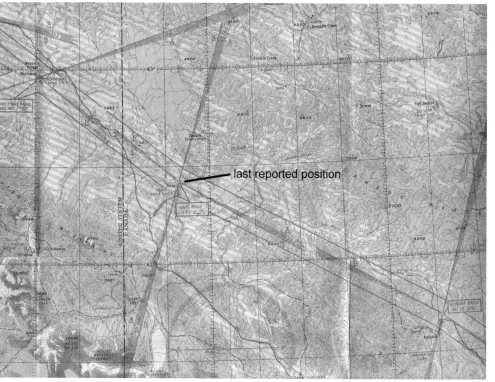

1950s World Aeronautical Chart showing the airway system flown by the lost C-54D.

on an almost daily basis, sending distress call after distress call without an answer, and watching the cold take their fellow survivors one by one. Trapped in a remote valley or hidden on a distant mountain, they never gave up hope for a rescue that never came.

Perhaps someday the mystery of the lost C-54 and its forty-four occupants will be solved. Somewhere in the remote wilderness of Canada the plane or what is left of it remains hidden, long since covered by snow, ice or a canopy of tall forest. Like other missing aircraft before it and since, it will eventually be found. Over time even the deepest snow melts, glaciers recede and forests are thinned.

Chapter 11—Without a Trace

Very little new information was uncovered over the ensuing decades that could explain the fate of the missing transport and its forty-four occupants. The search was reopened on two occasions during the spring of 1950 and the fall of 1959 in western Canada, after high-flying aircraft reported spotting the wreckage of a large four-engine plane. Although subsequent flights by several RCAF search aircraft could find no sign of previously unrecorded wreckage, there were old recorded crash sites of smaller WWII lend-lease and other military aircraft along the same route. Those well documented sites were determined to have been mistaken for the larger missing plane when viewed from a high altitude.

One of the most mysterious reported sightings of the missing C-54 occurred in July 1951 near Yakutat, Alaska. An aircraft involved in the search for a missing Canadian Pacific airliner reported sighting the wreckage of a large plane on Mt. Crillon, between Juneau and Yakutat.[2] The wrecked four-engine plane had USAF markings and was supposedly lying relatively intact in a small saddle below the summit of the mountain, with its engines and red colored wing tips clearly visible. At first it was believed to be the missing C-54 from January 1950, but was later identified as a different Air Force C-54 which crashed in July 1950. The site was initially misidentified as Mt. Crillon instead of Mt. La Perouse, located a few miles away.

A General Airways DC-3 was lost in the same area near Yakutat six months later.[3] Ironically, it was found on Mt. Crillon, where it had crashed with no survivors after flying off the airway. The wrecked fuselage had red markings and was identified as the missing General Airways DC-3 by one of the search aircraft. The next day, however, another plane which attempted to verify the sighting could find no evidence of a crash or wreckage.

It was believed an avalanche or heavy snow during the night must have hidden all evidence of the crashed DC-3. Whether search aircraft were in fact looking at entirely different mountains in an area with many similar peaks, or the wreckage was from one of many former missing military and civilian planes are questions that remain answered.

---

2 The missing Canadian Pacific Airlines DC-4 is discussed in detail in Chapter Thirteen.
3 Circumstances concerning the General Airways DC-3 crash are discussed in the author's previous book, *Broken Wings: Tragedy and Disaster in Alaska Civil Aviation*.

Chapter Twelve

February 13, 1950

Broken Arrow

A cold Arctic air mass had been inundating the interior of Alaska for nearly two weeks, plummeting temperatures into the negative fifties. Ice fog hovered over the local communities and airfields, reducing visibility to zero and delaying normal air operations. Ground crews were in a constant battle of attrition trying to keep aircraft airworthy, repairing and replacing broken components which frequently failed in the extreme temperatures. Aircraft that could not be stored in heated hangars had to be preheated before takeoff in order to prevent oil in the engines and hydraulic systems from freezing. Sensitive electronic equipment became fragile and wiring brittle to the touch after exposure to the harsh cold for too long a period. Even solid metal weakened by the numbing cold could fracture like glass under the worst conditions.

As harsh as the Arctic winter conditions could be in Alaska on men and equipment, they also provided excellent training opportunities for the military. Common sense dictated that if military aircraft could operate safely in Alaska during severe cold temperatures, especially the Air Force's long range strategic bombers, then they could also operate effectively in other Arctic environments of the world, even the coldest regions of the Soviet Union. Since the early days of World War Two the United States military had built and maintained several large airbases in Alaska with a strong contingent of aircraft. Those same airbases provided key support facilities and valuable training opportunities for other military units normally stationed outside of Alaska.

In February 1950 a squadron of new B-36Bs from Carswell Air Force Base in Fort Worth, Texas, was involved in a training exercise at Eielson Air Force Base near Fairbanks, Alaska. Designated as the 436th Bomb Squadron of the 7th Bomb

Group and assigned to the Eighth Air Force under the direct control of the Strategic Air Command (SAC), the bombers were a vital piece of America's nuclear counterstrike capability in case of war with the Soviet Union.

The winter training exercise required the bombers to deploy to Eielson AFB for refueling and servicing, where they would pick up their assigned combat crews and fly simulated nuclear strike missions against designated targets in the continental United Sates. Designed to test the operational capability of the squadron under realistic combat conditions and the effectiveness of long range deployments in adverse weather, the exercise was taken very seriously at all levels of command.

B-36B showing its distinctive rear mounted engines. (Brian Lockett)

Each B-36B bomber involved in the training missions would carry a MK IV "Fat Man" atomic bomb weighing more than five tons, almost identical in weight and appearance to the original "Fat Man" atomic bomb dropped on Nagasaki, Japan in August 1945. Except for the plutonium core which was removed during training missions, the bomb was authentic and still armed with several thousand pounds of conventional explosives and a small amount of unenriched uranium. The plutonium core was necessary for a fusion reaction to occur and without it the bomb was not a functional nuclear weapon. A replacement core was instead added as a replacement for the plutonium during training, duplicating the weight. Thirty-two electric detonators, installed for detonating the conventional explosives and to trigger a nuclear reaction with the plutonium core, were left in place during any training missions. This allowed the weapon and its components to be destroyed in case of an emergency.

Given the name Peacemaker by its designers, the B-36 became operational in 1948 as the world's first intercontinental bomber, capable of flying 6,000 nautical miles while carrying 10,000 pounds of payload. If necessary it could carry a maximum load of 86,000 pounds in the four internal bomb bays, but at a reduced range. Spanning a massive 230 feet from wingtip to wingtip and stretching 162 feet from nose to tail, the Convair B-36 remains the largest bomber ever built, surpassing the B-52 Stratofortress still in use with the United States Air Force. Six Pratt and Whitney R-4360-41 piston engines, mounted behind the wings, enabled the huge bomber to fly at more than 400 mph and up to a maximum ceiling of 40,000 feet. Defensive armament consisted of remote controlled 20mm cannons mounted in six retractable turrets and a fixed turret in the nose and tail.

B-36B departing Eielson AFB on a winter training mission, February 1954. (3rd Wing Historical Office, Elmendorf AFB)

During the late 1940s and early 50s the B-36 was the most advanced bomber in existence, but its service life was cut short by the introduction of jet aircraft which quickly surpassed the older technology in both capability and effectiveness. The B-36 was the last World War Two designed aircraft to see service in the U.S. military, and although later versions were modified with jet engines, they were only temporary improvements to an already outdated airframe. New jet aircraft designs and more modern technological advancements made the B-36 virtually obsolete by 1955, although several of the aircraft continued to serve in a reconnaissance role until 1959.

Ten aircrews from the 436th Bomb Squadron arrived in advance at Eielson AFB in early February, along with a full contingent of supply, maintenance and operations personnel. The B-36Bs were scheduled to be flown up from Texas by ferry crews a few days later, but the extreme cold and ice fog delayed their arrival for a week. Other problems occurred with the squadron's fuel trucks when the hoses and pumping units began freezing from prolonged exposure to the negative temperatures. Necessary repairs and procedures were eventually implemented as the airmen learned from their Alaskan counterparts how to operate effectively in the harsh conditions.

The first flight of B-36B bombers from Texas arrived at Eielson AFB on 11 February. Each aircraft was refueled, serviced and loaded for a quick turnaround, while the combat crews already on station received the last of their briefings on the simulated nuclear strike mission. Within a few hours the bombers were airborne again, flying the designated mission profiles for the exercise. The remaining squadron B-36Bs which arrived in Alaska over the next several days followed similar scenarios.

B-36 with the personnel and equipment necessary for routine support and maintenance. (Air Force Historical Research Agency)

The second flight of three B-36B bombers arrived the morning of 13 February, about the time their combat crews entered the Base Operations building. Each crew received a final exercise briefing to include detailed weather forecasts and mission procedures.

Assigned to the second aircraft on the mission were Captain Harold Barry and a crew of sixteen men. Flying in aircraft 075, their mission profile called for a cruise altitude of 12,000 feet between Eielson AFB and Seattle, following the airway system from Anchorage over the southeast coastline. After passing Seattle they would climb to 40,000 feet on a new route over Montana, before reversing course to the designated strike target of San Francisco. From there they would descend to 30,000 feet for the remainder of the flight, continuing over Nevada and the Caribbean before landing in Texas. The non-stop mission would cover approximately 6,000 miles and last sixteen hours.

Chapter 12—Broken Arrow

Captain Barry's crew consisted of two other pilots beside himself, Captain Theodore Scherier and First Lieutenant Raymond Whitfield, as well as bombardier First Lieutenant Holiel Ascol, navigator Captain William Phillips, flight engineers First Lieutenants Ernest Cox and Charles Pooler, radar operator First Lieutenant Paul Gerhart, radio operators Staff Sergeants James Ford and Vitale Trippodi, gunners Sergeants Dick Thrasher, Neal Staley, Elbert Pollard and Martin Stephens, and observers First Lieutenant Roy Darrah and Corporal Richard Schuler from the 7th Bomb Wing. Lieutenant Colonel Daniel McDonald, an evaluator from the Strategic Air Command, was only included to monitor the training exercise parameters of the flight.

Weather on the ground at Eielson AFB at takeoff consisted of ice fog and reduced visibility up through several hundred feet, but clear skies on climb out between Eielson and Anchorage. From Anchorage south to Seattle the weather was expected to deteriorate rapidly near a low pressure area situated in the Gulf of Alaska. A solid overcast was forecast between 1,000 and 15,000 feet, with scattered to broken layers above that as high as 20,000. Heavy clear icing was possible below 7,000 feet in clouds and precipitation, and heavy rime icing from 7,000 to 15,000. Turbulence was expected at all levels below 14,000 feet. Snow showers were reported from Anchorage southeast to Annette Island, near Ketchikan, turning to rain showers nearer Seattle. Winds aloft at their cruising altitude of 12,000 feet were forecast out of the northeast at 30 to 45 knots velocity. Because of the strong likelihood of encountering heavy icing conditions en route, flight below 17,000 feet was not recommended by the weather briefer.

During the mission briefing the Task Force Commander and 8th Air Force Operations Officer decided the three B-36Bs should remain at their assigned mission altitudes of 12,000 feet, based on personal assessments of the weather conditions and because a change in the mission parameters of the simulated war plan would cause unnecessary changes. The crews of each aircraft were instead instructed to climb above the clouds layers only if the heavy icing conditions were encountered. The first aircraft scheduled for departure was bomber 083, to be followed in approximate twenty minute intervals by bomber 075 and bomber 061.

After aircraft 075 arrived at Eielson AFB, maintenance crews began preparing the bomber for the simulated mission by correcting mechanical deficiencies noted by the ferry crew on the flight up from Texas. It was also refueled and serviced for the long flight ahead, taking more than 17,000 gallons of fuel. A total of slightly more than three hours was spent on the ground before it lifted off from the runway at 11:27 am. Of the three B-36Bs that landed earlier that morning, 075 was considered by maintenance personnel as being in the best condition.

The flight progressed normally after reaching its cruising altitude of 12,000 feet, continuing past Anchorage and turning southeast along the Alaska coast. A malfunction did occur with one of the communication radios, requiring the crew to relay position reports through the other B-36B flying ahead of them, but overall the aircraft was flying as expected. Several hours into the flight the expected low pressure area with its accompanying cloud layers and precipitation was encountered over the Gulf of Alaska, near Yakutat. In-flight icing conditions did not occur as forecast, however, until the bomber was approximately nine hours into the mission, south of Annette Island near the Alaska/Canada coastal boundary.

Consolidated B-36 with the side observation blisters visible near the nose and rear fuselage. (USAF via Howard Wynia)

There were no indications of trouble initially. Only light turbulence buffeted the aircraft when it first encountered snow showers, but that was soon followed by the sounds of what the crew thought was hail hitting the fuselage. That assumption quickly changed as the bomber's airspeed began decaying and all six engines began to over-speed; a likely indication of ice buildup on the propellers. The Flight Engineer reacted quickly by manually adjusting the propeller settings, and at the same time Captain Barry applied increased engine power in an attempt to climb above the icing conditions.

The waist gunners stationed on each side of the aircraft, who also served as spotters by watching the engines for possible malfunctions that would not normally be indicated on the cockpit instruments, reported the wings were beginning to collect a layer of ice behind the de-icing boots. By then ice was noticed by the flight engineer on an outside antenna, but the full extent of the icing could not be determined due to limited outside visibility. Night had already fallen and a thin layer of frost had coated many of the inside windows.

With the propeller over-speeds now under control the B-36B began a slow rate of climb for several minutes. Unfortunately, just when the crew thought they had overcome the emergency and relaxed again, the number one engine, outboard on the left wing of the aircraft, began indicating a high fuel flow with a corresponding drop in engine torque. The Flight Engineer immediately leaned the engine, momentarily correcting the problem, but a few minutes later all six engines began duplicating the same malfunction. Within seconds each propeller started surging uncontrollably. Airspeed and rate of climb began decreasing until the bomber was unable to maintain its altitude without full application of emergency power.

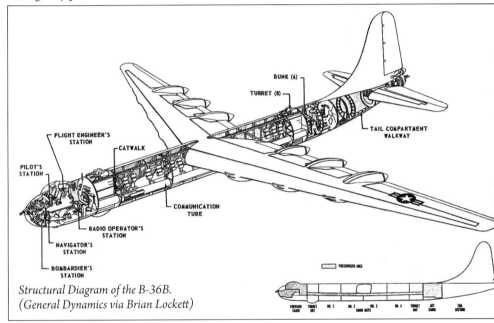

Structural Diagram of the B-36B. (General Dynamics via Brian Lockett)

As the crew struggled to hold the aircraft's altitude at 15,000 feet, the surging in the propellers intensified. Adjusting the propeller settings to a fixed position instead of automatic control helped reduce the problem but did not entirely alleviate the emergency. Captain Barry and his men were by then obviously concerned, but there was little they could do except try and maintain their altitude until exiting the icing conditions. Even though the situation was serious, it was not yet considered critical. Provided the engines kept running and maintained adequate power, the aircraft could continue in level flight. Icing on the wing and tail surfaces was also a concern due to the potential loss of lift and increased weight, but an excessive amount was being prevented, at least for the moment, by de-icing boots installed on the airfoils. Captain Barry knew if they could maintain a constant flight altitude long enough to fly out of the clouds, or burned enough gas to climb above them, the icing would be left behind and the mission still accomplished.

Whatever optimism the crew had, however, quickly changed after only a few minutes as the circumstances deteriorated even further. The left side gunner situated behind the port wing, reported a fire in the number one engine that was noticeably shooting flames around the edge of the cowling. Emergency steps to shut down the malfunctioning engine and feather the propeller effectively extinguished the fire, but with maximum power still applied on the remaining engines it was only a matter of time before they developed the same problem. Compelled by circumstance to power back the remaining five engines so they could continue functioning at reduced power, the bomber could no longer maintain altitude and began a slow descent. At this point it was already too late for the other engines, for within seconds ice that had accumulated in the carbu-

B-36B involved in Alaska cold weather testing. (General Dynamics via Brian Lockett)

retors triggered an increased fuel rate being metered into the systems until they became overloaded, leaking onto the hot engines.

Before the flight engineer could react to the excessive fuel flows by manually leaning the fuel mixtures, another fire occurred in the number two engine. Less than two minutes later the same thing occurred with the number five engine centered on the right wing. Both engines were immediately shut down and their propellers feathered, eliminating the fires, but with only three engines now providing power, the aircraft began descending at a much faster rate. Further thoughts of still continuing the training mission were forgotten as the crew focused on jettisoning the "Fat Man" nuclear bomb and saving themselves.

With no other options available Captain Barry turned the big bomber away from the coastal islands and waited until the radar operator verified they were over an open area of ocean. Only then did he order the bombardier to jettison the 10,000 pound bomb. It fell away offshore of Banks Island and was observed exploding

Chapter 12—Broken Arrow

3,000 feet above the water after being triggered by the automatic detonators. With the top secret weapon now destroyed and no possibility of the technology falling into the wrong hands, Captain Barry turned the stricken bomber back toward land. Having already lost 6,000 feet in altitude and still descending at more than 500 feet a minute, he had no choice but to apply emergency power to the three remaining engines in order to provide enough time for the crew to bail out.

Even with the lumbering bomber's rate of descent lessened by the application of emergency power, it dropped another 4,000 feet before nearing land. High coastal mountains ahead and the cold waters of the Pacific behind them, left the crew with no choice but to jump over the heavily forested islands along the Canadian coast. Nine and a half hours had passed since departing from Eielson AFB and only a half hour since first encountering the icing conditions which disabled their aircraft. Now they were bailing out into a biting snow storm and heavy cloud cover over a remote stretch of the British Columbia coast. A final distress call was transmitted only seconds before the crew began jumping clear of the aircraft.

Captain Barry was one of the last to leave after activating the autopilot that would steer the bomber on a southwest heading back out to sea. Only one other crew member was still on board by then, adjusting his parachute as Captain Barry moved toward the exit hatch. Captain Scherier, one of the three pilots, made eye contact and signaled he would be right behind him.

High winds dispersed the stricken bomber's crew over several miles as they were carried earthward. Most of the men came down on Princess Royal Island in a heavy forest of spruce and hemlock, causing several injuries. Other crew members were never found and were presumed to have landed offshore and perished.

Bomber 083 was flying twenty minutes ahead and 061 approximately thirty minutes behind when 075 announced it was experiencing multiple in-flight emergencies. The crews of 083 and 061 could only listen in frustration as Captain Barry notified them of their intention to bail out. There was little the crew of 083 or 061 could do to assist, except hope for the best. Luckily neither experienced the icing conditions which disabled 075, due to the fact the pilot of 083 changed his assigned altitude to 17,500 feet before departing Eielson AFB, and the pilot of 061 climbed above the clouds along the coast once Captain Barry announced his aircraft was in trouble. Unlike the crew of aircraft 075, the pilot-in-command of 083 was concerned about the forecast weather conditions enough to request an altitude change before takeoff. Both 083 and 061 were able to safely fly above the severe weather and proceeded without any problems. Aircraft 083 remained in the area for more than an hour, trying to contact the crew of 075 after they bailed out, without success.

Regardless of whether the plutonium core had been removed from the bomb, the device still had top secret nuclear technology incorporated in its design. As such the possession of the weapon or any loss of control of the weapon was a matter of national security. Once bomber 075 was reported as lost, a "Broken Arrow" emergency was declared by the Air Force. The code phrase "Broken Ar-

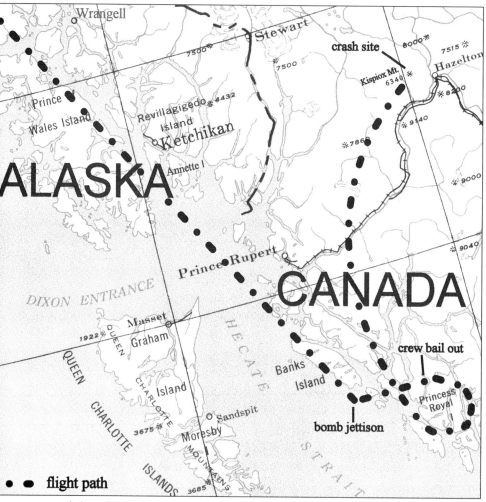

Flight path of the lost B-36B over Alaska and Canada.

row" was and still is used by the United States military to alert government agencies and security forces of an accident involving a nuclear weapon. If possible, any reported loss and ensuing search for a missing weapon is kept under a veil of absolute secrecy to prevent foreign governments or terrorists from acquiring the nuclear capability. When 075's crew jettisoned their 10,000 pound "Fat Man" bomb and disappeared with their aircraft over a remote region of wilderness, the accident became the first "Broken Arrow" incident of the Cold War. The true circumstances of the event were not revealed for decades.

A massive air, land and sea search ensued within hours of bomber 075's last distress call. By the following morning twelve Air Force B-29s were en route from bases in California, Washington and Montana, while fourteen U.S. Coast Guard aircraft and helicopters from the Royal Canadian Air Force began combing the area at first light. By then several Canadian and American ships dispatched from nearby ports had already arrived and began searching the surrounding waters and shorelines of the coastal islands where the missing B-36B was last reported. Since the search area was in the Canadian province of British Columbia, operational control was under the jurisdiction of the Royal Canadian Air Force.

By the afternoon of 15 February, nine of the abandoned bomber's crew who landed safely on the island reached the shore where they were picked-up by local fishing boats. A tenth crewman who sustained severe injuries was left in a makeshift shelter until other members of the crew could return with rescue personnel. He was found later that day by a search party from a Canadian destroyer.

No wreckage from the bomber or signs of the other seven men was located until the following day. Two more survivors were found on Princess Royal Island by a second search team from the Canadian destroyer *Cayuga*. Logic dictated the remainder of the missing crew would be found in the same general location, motivating the *Cayuga* and U.S. Coast Guard cutter *Wyonna* to send out additional search teams to comb the island. They were unsuccessful and subsequent searches of the nearby islands by other Coast Guard ground teams failed to find any trace of the five missing men.

While the hunt continued on ground and sea for the missing crew members and abandoned bomber, the survivors were carefully isolated from the media and transported back to the United States. Only after they had been debriefed was it learned the nuclear weapon on board had been safely jettisoned and destroyed before bail out. Even then the search continued in an attempt to find the plane's wreckage and hopefully determine what caused the engine malfunctions. Since the B-36 bomber fleet comprised a majority of the U.S. Air Force's nuclear strike capability and was a major deterrent to war with the Soviet Union, it was essential to analyze what went wrong and what corrective action was necessary to prevent the same occurrences in other aircraft.

Over the next few days additional military aircraft from U.S. and Canadian bases joined the search operation, often battling inclement weather that remained a persistent presence. Low overcast clouds accompanied by heavy rain and snow showers and winds up to 70 miles an hour delayed search efforts for days at a time. Tragedy also struck an aircraft en route to join the search when it crashed on takeoff in Great Falls, Montana, killing eight of the crew.

By 20 February no signs of the missing crewmen or bomber had been found. More than a hundred and fifty individuals were involved in searching the coastal islands by foot alone, with no success. By the end of the operation more than 25,000 square miles had been covered by numerous aircraft and ships.

A one-man survival raft was found in Hecate Strait near Princess Royal Island on 22 February, but there was no indication it had been used. Since each crewman had as part of his personal survival gear a small life raft, it was believed the one that was found was either torn loose during bail out and blown offshore, or had been inflated when one of the crew came down in the water. If one of the missing men had landed in the water, it was likely the cold temperature would have quickly weakened them to the point they were unable to climb into the raft or swim safety to shore.

All the members of the crew had been wearing heavy cold weather Arctic clothing that was essential for the harsh winters of Interior Alaska, but was unsuitable for use in the wet environment along the southeast coast. The five large survival kits carried aboard the aircraft, which provided several weeks of rations, signaling devices and other emergency equipment, were not dropped during the crew's quick exit from the plane. Whether or not the five missing crewmen survived their parachute fall is unknown, but if they survived the landing, a lack of adequate survival equipment would have certainly contributed to their demise.

Four of the five missing men were the first to exit the bomber as it approached land. They included Lieutenant Ascol, Captain Phillips, Sergeant Straley and Sergeant Pollard. The fifth missing man was Captain Scherier, the last person still on board when Captain Barry jumped. The surviving twelve crewmembers were found within three miles of each other, some on the shore of Princess Royal Island or the adjacent Ashdown Island. The men who came down closest to shore had exited the aircraft immediately behind four of the missing, indicating those four had most likely gone down in the ocean west of the islands. Their bodies were never found. Captain Scherier should have been the last to jump from the plane. His body was also never found and was presumed to have gone down in the waters on the opposite side of Princess Royal Island.

As the days progressed into weeks with no evidence of the missing men or aircraft revealed, the search efforts gradually scaled back and eventually halted entirely. For years the fate of the lost bomber remained a mystery.

Captain Barry along with other members of the crew provided a detailed account of what happened during the flight on 13 February to a Special Investigation Board. He was unfortunately killed a year later in another B-36 accident

Chapter 12—Broken Arrow

during a mid-air collision. One man who survived the mid-air had also served with Captain Barry aboard 075.

The Air Force Special Investigation Board convened on 16 February, three days after the accident. All aspects of the investigation were classified as Top Secret due to the sensitive nature of the bomber's mission. The report was submitted directly to General Curtis LeMay, the Commanding General of the Strategic Air Command (SAC). It was not declassified and made available until decades later through the Freedom of Information Act.

The findings of the Investigation Board were noteworthy for what they both did and did not determine. The conduct and action of the crew of 075 during the progression of the mission were found to be non contributing factors in the accident. In fact the crew was one of the most highly trained and experienced of all the B-36B crews in the 7th Bomb Wing. Captain Barry had more than 2,400 hours of flight time and was regarded as a strong, capable pilot. The other pilots, navigator, radar operator and three flight engineers on board were also rated as superior or excellent during previous evaluations.

A review of the bomber's maintenance records found no history of prior mechanical deficiencies with the engines which could explain the malfunctions. The aircraft itself was relatively new, as were all the B-36Bs in the inventory. Four of the six engines had each been flown 185 hours, while the other two had a lower total from being installed shortly before deployment to Alaska.

During the investigation it was revealed past incidents of engine fires on other B-36Bs in the Bomb Group had occurred and were attributed to problems with the exhaust system. One of the flaws was in the design of the airframe, which mounted the engines behind instead of forward of the wing, preventing adequate cooling from the ram air that was inherent on forward mounted engines. New exhaust systems designed to alleviate the problem had already been put in production months earlier and were being installed on the B-36B fleet as they became available. Unfortunately it was a slow process and bomber 075 still had the old exhaust system on five of its six engines, including the three which caught fire after encountering the severe icing.

Initial indications pointed to the exhaust systems as the cause of the accident until statements from the crew of 075 and the crew of another B-36B in the same area two days before led to a different conclusion. Each aircraft only experienced its multiple engine malfunctions after entering severe icing conditions, which indicated the carburetors were the probable cause of the fires and not the exhaust system as first surmised. During the few previous deployments of the B-36 bomber fleet, none had encountered severe icing that was present during the cold weather exercise in Alaska and crews were thus unaware of potential problems which could occur. Since the B-36B was a relatively new aircraft, many of its inherent design problems had still not been identified, much less corrected.

Ironically the accident with 075 could have been prevented if the circumstances of another squadron B-36B had been properly reported. That bomber was also involved in the same mission profile two days previously, flying an identical route, at the same altitude, with almost the exact weather conditions that later existed for 075. It was the first B-36B to depart Eielson during the exercise and like 075 proceeded normally until encountering severe icing conditions along the Canadian coast. Shortly thereafter three of the six engines malfunctioned, including two catching on fire. Both engine fires were extinguished, but one engine had to be shut down completely and the other could only operate on partial power. Cruise flight could not be maintained and the aircraft lost indications on several flight instruments when the pitot-static system became fouled with ice. Flying in and out of the icing conditions, the bomber began a series of slow climbs and descents while the crew applied different power settings on the engines. The disabled plane eventually burned enough fuel to climb into clear skies and was able to continue, but during several of the descents the crew had been close to bailing out.

A fourth engine lost power north of Seattle causing the crew to initiate an immediate landing at McChord AFB. During the descent back through ice laden clouds the remaining engines experienced the same irregularity, leaving the aircraft with insufficient power to maneuver. The bomber was luckily in position for a straight in emergency approach to the runway and was able to land without further damage.

Although the circumstances of the near disaster were reported to higher levels of command, for whatever reason the engine problems were not fully briefed to the crews waiting in Alaska. In fact the information they did receive only mentioned the aircraft experienced minor icing difficulty and a partial loss of instrument indications.

Carburetor icing on the first B-36B was determined to be the cause of the engine malfunctions. Both engines which caught fire had the older style exhaust systems installed and were found to have exhaust leaks during subsequent maintenance checks. The carburetor icing was believed to be partially caused by their position near the cooling fan on the engine, directly in line with the fan's discharged air.

The Special Investigation Board for 075 reached the same conclusion in regard to that aircraft's engine malfunctions. The design and position of the engine carburetors did not allow sufficient application of carburetor heat, causing ice to develop in the internal chambers, venturi and mixture control valves, when operating in moisture laden air and temperatures conducive to the development of icing. Formation of ice in the carburetor then triggered a higher fuel flow or rich mixture which could not be adequately corrected by manual control. Fuel that was being blocked by the ice then overflowed into the exhaust system, where it combined with the hot exhaust gases to cause a fire in the engine.

Following the investigation, the Board recommended more efficient carburetors be installed on B-36B engines or that the existing carburetors be modified with better carburetor heat capability. The recommendation was implemented and in the interim all B-36 aircraft were restricted from flying in instrument conditions until a new design could be installed. As a result of the investigation the improved exhaust systems in production were made readily available and installed on the remaining bombers. New operating procedures were also established for dealing with carburetor icing and an engineering section was established in the Strategic Air Command to retest all B-36 equipment currently in use or in production.

The fate of bomber 075 after its crew bailed out remained a mystery for almost four years. During that time no wreckage was found and it was presumed to have crashed and sunk somewhere along the continental shelf off the Canadian coast. When the crew jumped from the aircraft it was at approximately 3,000 feet and still descending on a southwest course away from land.

In September 1953 a RCAF plane involved in the search for a small civilian plane unexpectedly sighted the wreckage of a large military bomber on a remote mountain in British Columbia. At first the pilot thought he had stumbled upon the missing USAF C-54 transport which vanished with forty-four people aboard in January 1950, but after closer examination he could distinguish a much larger fuselage with mounts for six rear facing engines. The tail number and markings soon identified it as a missing USAF B-36B bomber from a flight in February 1950. Aircraft 075 had finally been found.

Somehow the bomber had crashed 240 nautical miles north of its last known position at an altitude 3,000 feet higher than when the crew bailed out near Princess Royal Island. The stricken bomber mysteriously managed to regain engine power on its own, climb to a higher altitude and fly inland for more than an hour through towering mountains as high as 9,000 feet. This was especially remarkable when you consider the autopilot had been set to fly the plane on a southwest heading out over the Pacific. Did Captain Scherier change his mind and stay with the plane after the rest of the crew bailed out? If he did, why head inland toward an even more remote area of wilderness instead of continuing along the coast and not attempt contact with the other aircraft?

The discovery of the lost bomber under such strange circumstances induced more questions than answers. Although it had survived the crash fairly intact, there was no evidence revealing how it arrived at its final resting place.

The broken, partially burned fuselage of the bomber rested below a rock strewn ridge at the 6,000 foot level of Kispiox Mountain, roughly 235 nautical miles west of Prince George, British Columbia and 115 miles east of Ketchikan, Alaska. It appeared to have crashed while in a shallow left turn, before the momentum carried it forward and down onto the steep side of a northwest facing, shale moraine. The tail, left wing and three port engines remained on the ridge where they had been torn from the fuselage, while the center and forward

Dispersed wreckage from the lost B-36 on Kispiox Mountain, British Columbia. (Doug Davidge)

sections supporting the right wing and engines stopped on a steep slide several hundred feet below. Pieces of equipment, personal items abandoned by the crew before bailout and ammunition from the 20mm cannons lay scattered around the interior of the plane.

A ground recovery team from the USAF attempted to reach the site on foot in late September, but they were forced to turn back by the harsh terrain and weather conditions brought on by the onset of winter. Additional ground teams tried again in late October, but their attempts were also suspended after finding the heavy snowfall and steep, icy slopes too dangerous for winter travel.

In the spring of 1954 a joint USAF and Canadian recovery team finally arrived at the crash site by helicopter. No evidence of human remains was reportedly found, leaving the fate of Captain Scherier and the details of how the bomber arrived there a mystery.

More than a week was spent searching the area, as well as identifying and

Chapter 12—Broken Arrow

A section of the B-36B's destroyed fuselage. (Doug Davidge)

One of the plane's wing sections at the crash site on Kispiox Mountain. (Doug Davidge)

USAF insignia on a piece of the wrecked fuselage. (Doug Davidge)

analyzing every piece of wreckage. All the top secret electronic equipment was salvaged from the wreckage before it was rigged with high explosives and detonated, ensuring any sensitive information on the B-36's capabilities would never fall into the wrong hands. The cockpit and fuselage forward of the wings was almost completely destroyed by the explosions and resulting fire. Some pieces of debris were thrown as far as 1,500 feet by the series of detonations, but a large section of the fuselage behind the wings remained intact. It contained nothing of value and was left untouched.

Another group from the Geological Survey of Canada stumbled upon the crash site while conducting a scientific survey of the area in 1956. They found a Geiger counter device in the wreckage and unexploded pieces of ordnance, including small arms ammunition, larger 20mm shells and electric detonators. Since the team did not know the circumstances of the crash at the time, it was virtually forgotten.

Scattered wreckage viewed from the top of Kispiox Mountain. (Doug Davidge)

Forty years later one of the individuals from the geological survey team approached a Canadian environmental group about the wreckage over concerns the site might contain radioactive material. His concern was raised by newly released information obtained through the Freedom of Information Act, describing the bomber's secret payload and resulting "Broken Arrow" incident. Rumors had already begun circulating on the Internet and among different environmental groups that the weapon carried aboard the B-36B bomber did contain the actual plutonium core, making it a fully functional device capable of an atomic explosion. Additional stories began circulating on whether the "Fat Man" bomb was actually dropped over the Pacific before the crew bailed out, as claimed by the Air Force, or remained in the bomber until it crashed on the mountain.

The concerns expressed by the former geologist resulted in another expedition to the crash site in August 1997, sponsored by the Canadian Department of National Defense and the Environmental Protection Branch of Environment Canada. Their intention was to conduct a radiological survey of the wreckage and surrounding area to verify if any radioactive contamination existed. No unusual radioactive readings in the wreckage or indications of surface contamination at the site were found. The Geiger counter and explosives found at the site

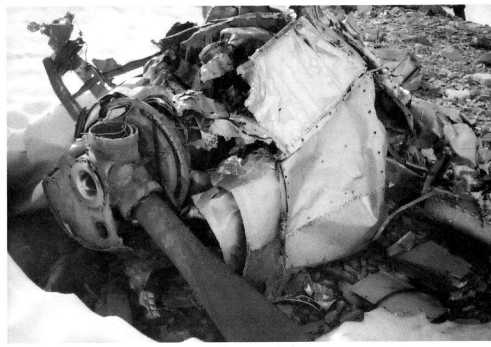

One of the six engines from the crashed B-36B. (Doug Davidge)

in 1956 were believed to have been left by the first recovery team in 1954 or to have been aboard the aircraft when it crashed.

Today there are still people who believe the bomber was carrying an active nuclear device. Decades of secrecy following the aircraft accident are in large part responsible for the reputed cover-up and near nuclear disaster. Some argue the nuclear weapon was active when it was dropped over the Pacific and is still lying on the bottom of the ocean, slowly corroding and unleashing deadly radioactive material. Others claim the plutonium core was carried in a separate container aboard the bomber and remained with the plane until it was recovered by the U.S. Air Force in 1954.

In 1950 the United States only possessed a few dozen atomic bombs and those were closely controlled by the Atomic Energy Commission. They could only be released to the Air Force on a direct order by the President when an actual war was imminent. Special Investigation Board documents and copies of statements from the crew of aircraft 075 confirm the bomb was not a live nuclear device and was jettisoned and detonated shortly before they bailed out. The official accident report, however, strangely omitted what happened to the plutonium core after being removed from the weapon before the mission.

What happened to Captain Sherier and how the bomber somehow regained power, turned and climbed above the mountains before crashing 240 miles from the crew's bail out point is a big part of the mystery. After the first recovery team returned from the wreck site in 1954 rumors began circulating that a skeleton and lead canister containing the plutonium core were found in the wreckage. If true, it only answers some of the questions and stimulates new rumors of possible espionage and conspiracy; making accusations of nuclear material aboard the bomber seem more plausible.

If perception is based in part on reality, then perhaps there is some truth to the stories. Until every remaining classified document, accusation and rumor is adequately addressed, the United States' first Broken Arrow incident will stay surrounded in controversy.

Chapter Thirteen

July 20, 1951

Cathedrals of Ice

The four powerful Pratt and Whitney radial engines hummed with efficiency as the Canadian Pacific Airlines DC-4 continued past the Alaska coastal village of Sitka. Heading north over the Gulf of Alaska at an altitude of 9,000 feet, the modern transport flew smoothly on a course to Elmendorf Air Force Base, near the territory's largest city of Anchorage. Most of the thirty-one passengers had finally nodded off to sleep after consuming their evening meal, allowing the two stewardesses to take a deserved break in the crew cabin located forward of the passenger area. Two experienced pilots occupied the cockpit, while the navigator, flight engineer and radio operator made up the rest of the crew at their assigned stations on the flight deck directly behind the pilots. It was the on-duty crew's responsibility to constantly verify course, speed, altitude, engine indications and all checks necessary for the safe operation of the aircraft.

The local time was approximately 10:50 pm as they crossed the airway beacon over Sitka and turned outbound on the northwest leg of the airway. Slightly more than four hours had passed since departing from Vancouver, British Columbia. It was one of four weekly contract flights scheduled for the United States Military Air Transport Service during the Korean War.[1] A refueling stop was planned at Elmendorf Air Force Base and another in the Aleutian Islands at Shemya before switching crews and continuing on to Tokyo, Japan. All thirty-one passengers were active duty military except three civilians attached to the U.S. Army. Two of the servicemen were assigned to the Royal Canadian Navy, twenty-three to the U.S. Air Force and three to the U.S. Army.

---

[1] Between August 1950 and March 1955, Canadian Pacific Airlines flew 703 charter flights carrying approximately 40,000 servicemen.

The commercial air route from Vancouver to Elmendorf Air Force Base covered nearly 1,400 miles, following the airway system along the coastal islands of British Columbia and Southeast Alaska, until cutting across the Gulf of Alaska to Yakutat. From there it continued northwest past Cordova and Prince William Sound before ending in Anchorage. South and west of the airway lay the storm tossed seas of the North Pacific, while north and east lay the high peaks of the St. Elias Mountains. Stretching across the rugged landscape for more than two hundred miles in a jagged wall paralleling the coastline, the massive cathedrals of sculptured rock and ice encompassed immense glacial ice-fields and snow covered valleys. Towering above the horizon, they reached unseen into the clouded night sky, beckoning unwary travelers who ventured too far off course.

Canadian Pacific Airlines DC-4, CF-CUL, one of four aircraft acquired from Pan Am in 1951. (CPA via Chris Charland)

The flight had been relatively smooth for the first several hours, with only light rain being encountered near Sitka on the west coast of Baranof Island. After proceeding past the range station out over the Gulf of Alaska, the winds increased in strength from an occluded front forming a hundred and fifty miles offshore. A cold air mass overtaking the warmer air closer to land began generating wide spread rain showers and a heavy cloud cover from the surface up to 12,000 feet. Above their cruising altitude, ominous accumulations of cumulonimbus clouds were formed by the converging air masses, quickly becoming unstable where they remaining hidden in the thick overcast.

At first the turbulence was light in strength and intermittent, barely stirring the sleeping passengers reclining in their cushioned seats or leaning uncomfortably against a window. The jolts of rough air soon increased in force and duration, shaking the aircraft enough that the stewardesses hurried through the passenger cabin, ensuring everyone had their seatbelts securely fastened. Holding on to the aisle seats for stability every few seconds as they walked between the double rows on each side of the cabin, the stewardesses completed the routine check and reassured the small number of passengers still awake there was no reason for concern.

Chapter 13—Cathedrals of Ice

On the flight deck the rest of the crew barely gave the turbulence a second thought, except to pass boastful anecdotes among themselves about previous experiences in some remote corner of the world when turbulent conditions were much worse and the planes were always overloaded or underpowered, requiring the greatest of skill. They were all well-seasoned veterans accustomed to occasional bouts of bad weather, especially along the turbulent coast of Alaska. The DC-4 they were flying was built for durability and safety and could easily endure severe levels of turbulence if necessary, although none of them wished it upon themselves or the weaker stomached passengers. As far as they were concerned it was just another routine flight.

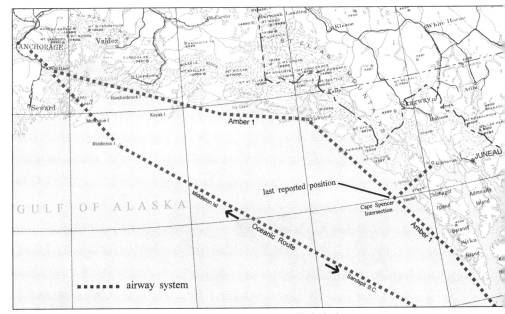

Coastal airway route and military offshore route across the Gulf of Alaska.

Approximately thirty minutes after passing Sitka at 11:17 pm, the crew made a routine position report crossing over Cape Spencer Intersection in the Gulf of Alaska, estimating their arrival over the Yakutat Radio Range at midnight. There was not mention of problems with either the aircraft or weather. At some point during the forty minutes it should have taken the plane to arrive over Yakutat, it vanished, never to be heard from again. No evidence of the Canadian Pacific Airlines DC-4 or its thirty-eight occupants has ever been found.

After the aircraft failed to report crossing the Yakutat Radio Range, contact was attempted by several communication outlets along the route of flight. When no response was received an emergency warning order was issued at 0:44 am to various search and rescue facilities, notifying them of the missing plane.

Coast Guard amphibian aircraft from Juneau were the first to respond by early morning and they were soon joined by a U.S. Air Force B-17 and C-47 already operating in the Juneau area on a training mission. A Coast Guard PBY Catalina from Kodiak also launched and began searching a route south from Cordova, in case the missing plane continued north past Yakutat before it disappeared.

Even though the summer sunlight provided ample hours of available daylight for search and rescue operations, poor weather conditions along the coast forced the search aircraft to remain below the clouds where effective coverage was limited by a low overcast, patchy fog and rain showers.

An additional five U.S. Air Force B-17s and a C-54 transport from bases in Anchorage and Fairbanks were en route by early afternoon, as were two SB-17s from McChord AFB, outside Seattle, specially configured for dropping small boats and survival equipment. Three Coast Guard cutters based in Juneau and Cordova were also dispatched to begin sweeping the coastal waters between Cape Spencer and Cape St. Elias on either side of Yakutat.

Twelve aircraft from the U.S. Coast Guard and Air Force and four from Canadian Pacific Airlines were involved by the following morning. Because of heavy clouds and precipitation which blanketed the search area, most were forced to stay in a small sector offshore where they were as busy staying out of the clouds as they were dodging each other. A few planes did climb above the thick cloud cover and were able to over-fly the tops of the higher coastal peaks, but nothing was sighted.

SB-17 search and rescue aircraft, parked and ready at Shemya airbase, Alaska. (Dan Lange)

There was no change in the weather until the fourth day of the search on 24 July, when clearing skies allowed twenty-seven aircraft to extensively cover the entire search grid for the first time. A wide area extending well offshore, as well as nearby coastal islands and dozens of inland mountains above 9,000 feet were systematically searched. Over-flights of the area continued through the night to better observe possible distress fires or flares, without success. Although favorable weather conditions persisted for several days, allowing search efforts to continue under the direction of the Coast Guard in Juneau, the absence of a single sighting of the missing aircraft left little hope anyone would be found alive. In spite of diminishing optimism, search crews retained a strong determination to at least find evidence of what happened and hopefully recover some or all of the bodies of the thirty-eight occupants.

The first sighting was made on 27 July by a plane searching the inland peaks of the towering St. Elias Mountains in the Fairweather Range, south of Yakutat. A fairly intact fuselage, four engines and wings painted with red tips were observed lying in a basin below the summit, partly covered by ice and snow at the 9,000 foot level of Mt. Crillon. Strangely, the report claimed the wreckage had identifiable U.S. Air Force markings and not the markings of a Canadian Pacific airliner, leading to speculation it was the missing U.S. Air Force C-54 which disappeared six months previously on a flight through western Canada.

Further investigation of the crash site, however, revealed the plane was not on Mt. Crillon but on Mt. La Perouse, and was the previously known wreckage of an Air Force C-54 transport which crashed a year earlier. The plane had flown off the airway system in bad weather while flying north along the coast, having last reported crossing Cape Spencer Intersection. The situation was eerily similar to that of the lost Canadian Pacific DC-4.

Overseas National Airways DC-4 being refueled before departure. (1000aircraftphotos.com)

A year and a half later and six months after the Canadian Pacific DC-4 disappeared, a cargo DC-3 belonging to General Airways fatally impacted Mt. Crillon under the same circumstances.[2]

As the search progressed through late July aircraft expanded their coverage to include a missing Norseman NY-4/5 carrying a well known Alaskan bush pilot. The single-engine aircraft piloted by Maurice King was flying in support of the Arctic Research Institute's study on glacial activities in the St. Elias Mountains. It vanished with two passengers after departing Seward Glacier, northwest of Yakutat, for a site near Mt. Hubbard, further east. Nothing was found, in spite of the intensive search activity in the area over the next few months.

By the end of July the overall operation extended further north and south

---

[2] The General Airways DC-3 crash is discussed in the author's previous book, *Broken Wings: Tragedy and Disaster in Alaska Civil Aviation*.

of the DC-4's last known position. Dozens of aircraft from military and civilian sources in Alaska, Canada and the west coast of the United States were still involved on a daily basis. While most of the larger multi-engine planes searched the offshore waters and high, rugged peaks of the St. Elias Mountains, smaller single-engine aircraft covered the coastal islands and inland waters south of Yakutat. Canadian Pacific Airlines stationed several of their planes in Yakutat for the search and were joined by other commercial air carriers from Juneau. Aircraft from as far as Whitehorse and Edmonton, Canada also participated in the operation, as well as Civil Air Patrol and private aircraft from local communities along the coast.

Alaska Airlines DC-4 after engine start. (Museum of Flight, Seattle, WA)

Another promising report occurred on 30 July from a range station controller in Whitehorse, who reported an unidentified aircraft requested a message be relayed to Elmendorf Air Force Base the same night the missing DC-4 disappeared. The message from the plane was received an hour after the DC-4's last position report over Cape Spencer Intersection and asked that Elmendorf be notified to have meals ready for the passengers upon their arrival. At the time of the communication there were no other passenger aircraft en route to Elmendorf AFB, leaving the missing DC-4 as the likely source of the radio call. If so, it also placed the plane well past Cape Spencer Intersection and north of Yakutat.

Once news of the radio message was received by search headquarters in Juneau, the operation began concentrating on the route segment between Yakutat and Anchorage. A dozen more aircraft from Elmendorf AFB were added to the search effort that now covered a larger area east and south of Anchorage into the coastal Chugach Mountains and Kenai Peninsula. One promising sighting reported seeing metal wreckage in the Kenai Mountains and another claimed seeing scattered debris on a remote peak on the north side of the St. Elias Mountains. Both were investigated and found to be only sunlight reflecting off chunks of ice.

The radio message from the unidentified passenger plane sent to the controller in Whitehorse also was later determined to be without merit, after the controller admitted being unsure of the exact day the message was heard. He claimed to have relayed the message onto Elmendorf, but the forwarded message was not received by any other station and therefore could not be confirmed as to what day it occurred.

While a large number of search planes were diverted north because of false sightings and unconfirmed reports, still others shifted more than two hundred miles south to cover Prince of Wales Island, northwest of Ketchikan. Several individuals in the community of Craig, on the outer coast of the island, came forward to report a large aircraft was heard flying overhead on the night the missing DC-4 disappeared. Although it seemed unlikely the plane would have turned around after passing Cape Spencer Intersection without radioing its intentions, the report was investigated thoroughly. Over-flights of the area and other islands surrounding Prince of Wales Island found nothing of relevance.

Within a few days the majority of aircraft were back operating near Yakutat after more potential pieces of evidence were discovered. The first was a yellow life raft that was sighted by an Alaska Coastal Airlines plane near the north end of Yakobi Island, less than thirty miles from where the missing DC-4 last reported. Part of the survival gear aboard included three survival rafts that would have been used after ditching in the water. A search plane and Coast Guard cutter sent into the area later that day found no trace of the raft or similar type object. The sighting was believed to be credible, but its origin could have been from any number of commercial fishing boats which operated in the area throughout the year as well.

About the same time as the sighting of the raft, a ground search team from the U.S. Coast Guard cutter *Storey* located a worn pair of trousers on a stretch of beach south of Yakutat, near the large decomposed remains of an apparent human leg and foot. A case of Army rations was also found further away where it had washed ashore with the tide. The bones were sent to Juneau for analysis and were initially identified by three local doctors as coming from a large human male. The amount of decomposition suggested the remains had been in the water for approximately two weeks.

Other medical personnel didn't agree and stated they appeared to be from a bear, which have similar bone structures. The trousers were later found to be too small for the large leg bone and were identified by an attached laundry tag and stitching as belonging to a fisherman who was lost in the Yakutat area the previous winter.

Several days passed before the remains were studied by forensic experts from the FBI in Washington and confirmed as being not of human origin, but from a bear. The case of Army rations was also dismissed as evidence from the missing plane when it was determined that particular type of emergency rations was not aboard the plane. No other potential clues materialized over the next month.

Heavy fog and low clouds moved over the search area during the first week of August, interfering with flights around Yakutat, but twenty-six aircraft remained involved in the operation until the search began scaling back a few days later. By

Northwest Airlines DC-4 unloading cargo. (Randy Acord)

the middle of the month the total number was reduced more and more as aircraft returned to their home bases or were diverted to other ongoing searches for missing aircraft elsewhere in Alaska and Canada. By the time the operation was officially halted on 31 October, thousands of flight hours covering tens of thousands of square miles were utilized in one of the largest aerial searches ever conducted at that time. Not a single piece of substantiated evidence was ever uncovered.

The Canadian Department of Transport investigated the incident and could find nothing conclusive indicating why the aircraft disappeared. It was determined to have sufficient fuel aboard for the flight and was not believed to have any major mechanical or structural problems which contributed to the disappearance. The log book and certification were carried onboard the aircraft, which was normal at the time.

The missing Canadian Airlines DC-4 originally entered service with the U.S. Army Air Forces during World War Two and was subsequently transferred to the U.S. Navy. It was later designated as war surplus and purchased for civilian operations by Pan American Airways. The company flew it under the name *The Clipper Winged Racer* before it and four other DC-4s were sold to Canadian Pacific Airlines in early 1951. After joining the Canadian Pacific fleet they replaced the similarly designed, but less economical Canadair Fours.

Canadian Pacific Airlines DC-4, CF-CUJ, waiting on passengers. One of four aircraft purchased from Pan Am in 1951. (Ed Coates)

In the Summary Accident Report issued by the Canadian Department of Transport, an error listed the total crew aboard as six, mistakenly omitting the second stewardess, Eva Lee. Other important facts revealed in accident investigation documents were also left off the Summary Report, including possible adverse weather phenomenon and the likelihood the navigator and radio operator were not at their assigned stations during the course of the flight.

In 1951 the high altitude, high speed flow of atmospheric winds known as the jet stream was only beginning to be understood. The strong band of westerly winds were at first believed to only occur in the upper troposphere, but pockets of the fast moving air were later observed diverting away from the upper stream in brief and inconsistent patterns as low as 8,000 feet. Royal Canadian Air Force navigators who flew a more direct offshore military route between Sandspit, British Columbia and Middleton Island, Alaska during the 1950s, reported incidents of extremely powerful, unforecast westerly winds being encountered at velocities more than 100 knots.³ In some instances course corrections as high as 90 degrees were required to keep the aircraft from being blown into the mountains. At other times the dangerous winds disappeared as quickly as they emerged. Any aircraft that was flying the closer-in, coastal airway and that experienced the high speed winds, especially in marginal weather conditions when signal reception was affected, would have been susceptible to extreme navigational error. That phenomenon might explain many of the unresolved aircraft disappearances in the area.

Canadian Pacific Airlines did not require the navigator and radio operator to be present on the flight deck during the seven hour flight from Vancouver to Anchorage, due to the number of radio range facilities and communication outlets available along the coastal route. Possibly they were not at their flight stations at

Canadian Pacific Airlines DC-4, CF-CUK. One of three sister aircraft of the missing DC-4 acquired from Pan Am. (Ed Coates)

the time of the disappearance, but instead were resting in the crew cabin for the next leg of flight between Anchorage and Shemya.

All of the crew was experienced in instrument flight procedures and DC-4 operations. Captain Victor Fox had logged more than 10,000 hours of flight time, including seventy hours in the previous thirty days. He was hired by Canadian Pacific Airlines in 1942 after many years of flying with smaller regional airlines. First Officer Bruce Thomson was a former airline captain with a smaller overseas company and similarly qualified with more than 6,000 total hours in the air. He joined Canadian Pacific Airlines in 1949 as a fully qualified navigator before being later promoted to First Officer. Both pilots had a previous history of poor radio range flying techniques during check rides and evaluations given by Canadian Pacific Airlines and the Canadian Department of Transport. Those

---

3 The military route was 90 nautical miles further west of the commercial route using Cape Spencer Intersection.

earlier deficiencies in instrument flight, however, were corrected preceding their assignments on the North Pacific route.

The navigator, Second Officer E.L. Krausher had not yet received a Canadian Department of Transport Navigator's license, but had held a prior First Class license from the British government, which expired the month before the plane vanished. He did have nine hundred hours of navigator time credited, of which two hundred and fifty were logged during the previous three months. In addition to being without a valid license, he had not been flight-checked by the company's chief navigator per standard policy. There was no information in the accident report on whether he had flown the same route before.

Flight Engineer Arthur Boon, Radio Operator Fred Tupper and Stewardesses Kathleen Moran and Eva Lee made up the rest of the crew. Ms. Moran was not

Northwest Airlines DC-4 taxiing for takeoff. (Museum of Flight, Seattle, WA)

initially scheduled on the flight, but had changed places with another stewardess so she could be fitted for a wedding dress in Tokyo. A third stewardess assigned to the flight had been taken off at the last minute to cover for a sick co-worker on a different aircraft.

A detailed weather briefing was received by the pilots and navigator before departure from Vancouver, which detailed the weather conditions expected along the route and over the Gulf of Alaska. Subsequent analysis of the weather patterns that existed at the time of the disappearance showed the briefing was fairly accurate, with the exception of embedded cumulonimbus clouds and higher than anticipated winds from the southeast. Cumulonimbus clouds are associated with thunderstorms and unstable air currents that can often cause severe turbulence and violent wind shear, as well as heavy rain or snow activity. At the time of the flight the freezing level was forecast at 9,000 feet, but following the disappearance was determined to be as low as 7,000, increasing the chance of severe icing being encountered.

Possible navigational error or radio interference was not mentioned in the Summary Accident Report, most likely because those assumptions could not be substantiated. Flight checks of the three radio range systems at Sitka, Gustavus and Yakutat were later conducted by the U.S. Civil Aeronautics Administration, which determined they were operating normally. Since the flight checks were performed in fair weather conditions several days after the disappearance, the possibility of interference from precipitation static or a deflection of the radio beam at the time of the disappearance cannot be dismissed. During the early years of instrument airway navigation, every low-frequency radio range had its own particular peculiarities that were magnified in the worst weather when pilots relied on them most. The radio range signal from Yakutat had been reported as bending or swinging inland away from the normal airway course on previous occasions.

A Loran (long-range navigation) system was installed on the missing DC-4, but was only accessible from the navigator's station. When operating correctly it determined the aircraft's position from a measured difference in electronic pulse signals transmitted by fixed ground stations. Susceptible to interference from precipitation and cloud cover, it was often useless in poor weather conditions, much like the low-frequency radio range systems.

A Northwest Airlines DC-4 flying south along the airway at 10,000 feet during the same approximate time as the missing Canadian Pacific DC-4, reported encountering lightning, much stronger than forecast winds and icing. Even more significant was the pilot's report of poor radio reception and considerable erratic swinging of the Yakutat course signal.

After the Canadian Pacific Airlines DC-4 departed Vancouver at 6:42 pm, it proceeded normally to the Sitka Radio Range without incident. A mandatory position report upon passing the Sitka Range was sent twenty-three minutes late, possibly because radio contact could not be established with the station controller. The report was instead passed to Yakutat, as was the next and last position report over Cape Spencer Intersection. The crew gave no indication of poor radio reception or problems with the aircraft, indicating the delay could have been caused by confusion over the aircraft's position. If the radio range information being received at the time did not coincide with what was anticipated, an ensuing delay would have occurred while the crew attempted to interpret and verify the conflicting information.

A two degree course change in the airway bearing occurred when passing over Cape Spencer Intersection at the mid-point between the Sitka and Yakutat Radio Ranges. At that point the crew would change the radio frequency, switching navigation from outbound on the northwest leg of the Sitka Range, to inbound on the southeast leg of the Yakutat Range. Both range stations operated on similar frequencies with similar Morse code identifiers.

Chapter 13—Cathedrals of Ice

All Radio Range stations transmitted an "A" and "N" Morse code signal for referencing positions on the quadrant legs, or airways. These signals were heard as a series of dots and dashes through the headphones. When flying outbound the "A" quadrant was on the left and the "N" quadrant on the right. Inbound signals were always in reverse, with the "A" on the right and "N" on the left. Depending on what quadrant signal was being heard and whether the aircraft was flying inbound or outbound, a position in relation to the airway centerline could be determined.

Pilots flew outbound on the right side of the airway from Sitka inside the "N" quadrant, and after crossing the intersection would stay on the right side of the

Northwest Airlines DC-4 over the midwest United States on a fair weather day. (Museum of Flight, Seattle, WA)

airway using the "A" signal inbound to Yakutat. However, if the radio frequency was not tuned to Yakutat after crossing the intersection, the pilots would have continued receiving the "N" quadrant signal, giving a false impression they were flying left of the airway. Attempting to correct the perceived error by steering further and further to the right, they could have easily flown into the high peaks of the coastal mountains.

Following the fatal crash of the General Airways DC-3 on Mt. Crillon in 1952, the similar frequencies and station identifiers at Sitka and Yakutat were changed.

Although mechanical or structural failure cannot be ruled out as a cause of the loss, a navigational error would seem more plausible. The three incidents involving the Air Force C-54 in 1950, Canadian Pacific DC-4 in 1951 and General Airways DC-3 in 1952 occurred under too similar of circumstances to be only coincidence. In each case poor weather conditions were prevalent over the Gulf of Alaska, each

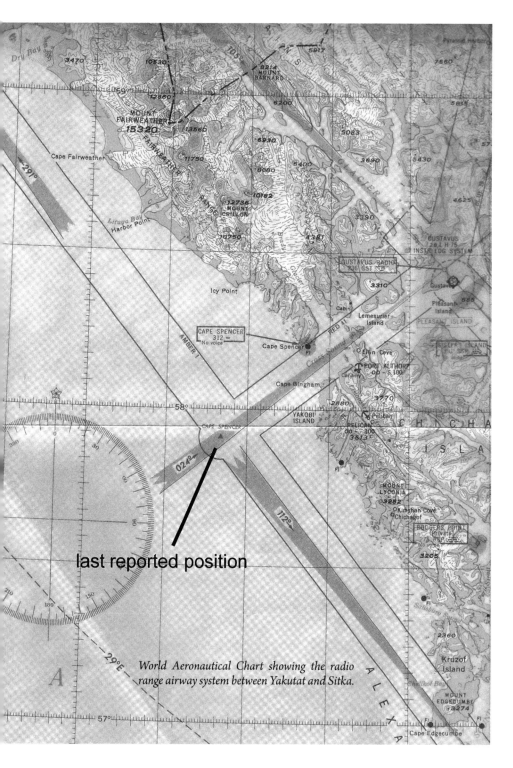

last reported position

World Aeronautical Chart showing the radio range airway system between Yakutat and Sitka.

was proceeding north on the Amber One airway at 9,000 feet and all made their last position report over Cape Spencer Intersection. The Air Force transport crashed into Mt. La Perouse, twenty-four miles north of the airway and approximately fifteen degrees off course, while the General Airways plane impacted Mt. Crillon, twenty-eight miles north of the airway and sixteen degrees off course. Only seven miles of separation existed between the two mountains.

The disappearance of a Pacific Alaska Air Express DC-3 in 1948 was also remarkably similar, with the exception it was flying south on the airway at 10,000 feet. It too vanished after reporting over Cape Spencer Intersection in poor weather conditions. Many other military and civilian aircraft have been lost in the area since World War Two, some while flying the airway system and others while operating under visual flight conditions. Almost all of have never been found.

For several days following the disappearance of the Canadian Pacific DC-4, the coastal mountains could not be adequately searched due to heavy clouds blanketing the region. The delay could have allowed shifting snow or avalanches to cover any signs of wreckage. Within fifty miles of the coast along a hundred mile corridor between Yakutat and Mt. La Perouse alone, fourteen ice-chiseled peaks extend above the 9,000 foot level, any of which could have caught the unsuspecting aircraft if it strayed inland off the low frequency airway. The resulting impact could have hurled the wreckage thousands of feet down the mountain, where it was then covered by an avalanche or a layer of windblown snow. Dozens of other high mountain peaks above 9,000 feet were located north and east of Yakutat, awaiting any aircraft which blindly made it between the towering summits nearer the coast.

Most likely the remains of the lost Canadian Pacific DC-4 and its thirty-eight occupants are entombed on one of the many cathedrals of ice rising above the coastal landscape. Perhaps someday it will be found and the mystery revealed. Until then we can only wonder as to its true fate.

Chapter Fourteen

November 15, 1952

Operation Warm Wind

Strong easterly winds blew across Cook Inlet on the southern coast of Alaska, jolting the Air Force C-119C transport as it climbed away from Elmendorf Air Force Base and turned on a westerly intercept toward the airway intersection over Delta Island. A low overcast and heavy snow showers were encountered soon after departure, enveloping the aircraft in a blanket of white that obscured all visual reference with the surface.

C-119 being loaded with troops at Elmendorf, AFB, January 1955. (3rd Wing Historical Office, Elmendorf AFB)

Twenty minutes later a second C-119C departed over the same route, followed by three others using the same separation sequence on the airway system. All five aircraft and crews belonged to the 76th Troop Carrier Squadron of the 735th Troop Carrier Wing, based in Miami, Florida and on deployment at Elmendorf AFB, Alaska for a joint Army-Air Force training exercise. Known as operation "Warm Wind", the joint exercise was being conducted over a two week period and was designed to test the response and maneuver capabilities of various armed forces in the Alaskan Theater.

Aircraft 570 was the third C-119C in the flight to depart from Elmendorf AFB, outside Anchorage. It lifted off at 11:40 am local time on an instrument flight clearance to Kodiak Naval Air Station, located 230 nautical miles south on Ko-

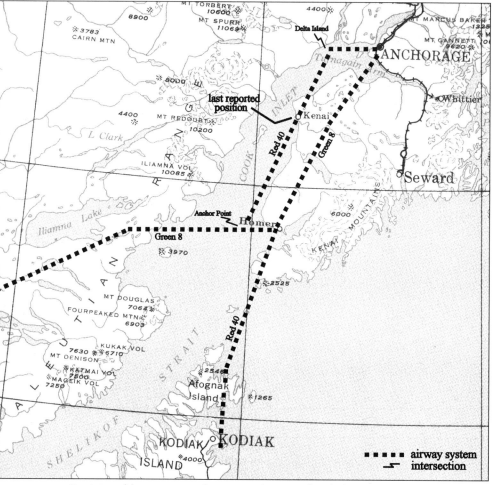

Depiction of the airway route between Anchorage and Kodiak.

diak Island. On board was a five man crew, fourteen Army soldiers from the 196th Regimental Combat Team at Ft. Richardson and an Air Force medic temporarily assigned to the team for the exercise.

Chalk three, as it was called, was cleared by Anchorage Air Traffic Control to climb and maintain 8,000 feet to Kodiak. The route followed Red Airway 40 to Delta Island Intersection, the Kenai Radio Range and Anchor Point Intersection, then east along the Green Airway 8 to Homer Radio Range, finally turning south again on Green

Airway 40 to Kodiak Radio Range. The route was identical to that flown by the previous two aircraft and later by the following two aircraft in the flight.

Twelve minutes after departure chalk three reported level at 8,000 feet, east of the Delta Island Intersection located 21 miles from the Merrill Radio Range. Communication was attempted with the Anchorage Approach Controller after takeoff, but radio contact could not be established and the pilot communicated directly with the Elmendorf tower controller from then on.

The next transmission was received at 11:58 after a five minute delay between the aircraft crossing the intersection and the pilot sending a position report. An estimated time of arrival of 12:07 was also sent for arrival at the Kenai Radio Range. A second delay occurred over Kenai when the pilot reported their crossing of the beacon ten minutes late at 12:16. Whether the delays were caused by radio difficulties or navigational problems in fixing their position is unclear, but no further contact with the flight was received before it vanished.

Statements from the pilots of other aircraft in the flight following the disappearance confirmed the in-flight weather encountered was basically as forecast. Overcast conditions up to 7,500 feet were predominate over Cook Inlet and the Kenai Peninsula at the time, changing to broken cloud layers south of Homer and scattered clouds over Kodiak. Moderate turbulence and light rime icing was present during the flight with slightly stronger easterly winds than were predicted at altitude. Because of the high winds pushing the aircraft right of course, pilots reported holding a wind correction of 35 to 50 degrees in order to maintain the desired track. All the pilots also reported difficulty in receiving the Kenai Radio Range because of precipitation static.

When aircraft 570 in chalk three failed to report over Anchor Point Intersection or Homer, numerous attempts at contact were made by various stations along the route. The chalk four aircraft, twenty minutes behind 570, also tried contacting the overdue aircraft several times, with no result. At 1:35 pm, approximately 30 minutes after chalk three should have arrived at Kodiak Naval Air Station, a communications alert was initiated. After all efforts failed from airfields and communication outlets in the region, it was declared missing and a search immediately initiated.

The search operation was organized under the operational control of the 10th Air Rescue Service Group at Elmendorf AFB. In addition to numerous aircraft being assigned to assist with the search, all available information was analyzed from various sources to determine where the missing aircraft might have crashed or ditched along its route of flight. A radar track of the route was immediately made available by the military radar facility at Elmendorf AFB.

During the time 570 departed the airbase and was en route to Kodiak, the radar equipment at Elmendorf was being tested for range accuracy after completion of a scheduled maintenance upgrade. Operators monitoring the radar scope tracked an unidentified aircraft flying outbound at 12:00 on a 240 degree heading from Elmendorf for several minutes. The radar plot showed the target continued across Cook Inlet at a 170 knot airspeed until four miles inland of the western shore, then turned on a 180 degree heading for another two minutes before the track faded and was lost from radar.

Fighter interceptors were scrambled from Elmendorf AFB to intercept the target as part of the "Warm Wind" exercise, but the radar plot was lost before they reached the area. At first it was assumed the plot was from one of many bush planes frequenting the area, since the target tracked across the inlet before turning south. The smaller local planes operating in visual flight conditions normally flew a similar route. Once the C-119C was declared missing, the plot was reanalyzed and determined from airspeed and position reports to be the same aircraft.

If the radar plot was in fact that of aircraft 570, it suggests the crew experienced navigational problems with identifying the airway intersection, overflew the intersection and then turned on a parallel course toward Kenai that was at least 12 miles west of the desired track. Since precipitation static was present at the time and position reports received from 570 were delayed by several minutes, they seem to substantiate the likelihood of navigational error.

Sketch of C-119 over Korea by aviation artist Jack Fellows. (Jack Fellows)

Even a parallel course 12 nautical miles from the airway would have kept the aircraft several miles east of the higher mountains in the Alaska Range. But if an inadequate wind correction was maintained during the flight toward Kenai, which was certainly possible if the station could not be properly tuned to track the signal, stronger than forecast winds at altitude could have pushed 570 into the higher terrain. For all that to happen would have required several cumulative errors in navigation beginning at the time of departure..

This seemed unlikely since the pilot-in-command, Captain Russell Peck, was an experienced instrument and instructor pilot with more than 4,600 flight hours. The co-pilot, Captain Edwin Boyd, was less experienced, however, and had only accumulated 47 hours in C-119 type aircraft. He was also not instrument qualified and had only been recently reinstated on flight status, which was why he was assigned to fly with an instructor pilot. In addition, a third pilot stationed at Elmendorf AFB and familiar with Alaska flight operations, was assigned to the flight by the Alaskan Air Command to preclude any navigational errors over the route, which was unfamiliar to the crew.

Captain Howard Marks was only aboard to assist with navigation, if necessary, and was not qualified in C-119 aircraft. He was fully instrument certified and his placement on the aircraft was the direct result of another C-119C from the 735th Troop Carrier Wing crashing into a mountain northwest of Anchorage the week

C-119 of the 314th Troop Carrier Group being refueled at Whitehorse, Yukon Territory, 1955. (USAF via A.T. Lloyd)

before. That crash was a direct result of navigational error. The aircraft was also involved in the "Warm Wind" exercise and all 19 occupants were killed on impact.

The other two crew members assigned to the missing flight a week later were the crew chief, Airman First Class Jimmie Robertson and the radio operator, Airman Second Class John Landis. All of the crew except Captain Marks were assigned to the 76th Troop Carrier Squadron out of Miami, Florida.

Immediately following the C-119s disappearance, search efforts were coordinated along the entire route from Elmendorf to Kodiak, concentrating around the area on either side of the airway system between Anchorage and Homer. Six aircraft from Kodiak NAS and eleven others from Elmendorf AFB were involved in actively searching the first few days, while five additional planes with rescue personnel were held on the ground until the missing C-119 could be located.

On 16 November several search planes focused on the Mt. Iliamna and Mt. Redoubt areas on the western side of Cook Inlet, where the radar track of the unidentified aircraft had been heading. Both mountains extended more than two thousand feet above the missing plane's flight altitude and were the closest to the course it had been flying. Although eleven recent snow slides were found in the area and investigated, all appeared to have resulted from heavy snow accumulations and not by the impact of a plane.

Low clouds, snow showers and decreased visibility over the search area limited the coverage on 17 and 18 November, but all the planes were flying as much as the weather permitted. Often the aircraft would fly until nightfall in the areas of good weather then return home and land in terrible conditions. Often push-

Fairchild C-119 over a remote stretch of coastline. (USAF via Howard Wynia)

ing their physical limits to the point of exhaustion and stretching the endurance of the machines as far as they could, it was a credit to the airmen and ground crews that no search aircraft were lost during the operation.

By the 19th the weather finally lifted and allowed a detailed search under clear skies for the first time. Aircraft began concentrating around the Kachemak Bay area near Homer where some possible sightings and signals from the missing aircraft were reported, but all were eventually found to be without merit.

On the following day heavy fog settled over the coast and offshore waters, obscuring the entire search area from Anchorage to Kodiak and effectively grounding the aircraft for most of the day. Even those able to depart later that afternoon and evening were forced to fly on instruments with limited visual references for the majority of the flight. Several of the planes maintained a listening watch on the emergency frequency, trying to verify another possible radio signal that had been received the day before, but they were unsuccessful.

Chapter 14—Operation Warm Wind

Clear weather conditions on 21 November allowed extensive over-flights of the search area by larger aircraft, but high winds kept most of the smaller planes and helicopters grounded. The high winds also damaged some of the search planes parked at Elmendorf AFB, making them unusable for several days.

RCAF C-119s at Cold Lake, Alberta. (1000aircraftphotos.com)

Thirty C-119s of the 314th Troop Carrier Group deployed at Elmendorf AFB, 1955. (USAF via A.T. Lloyd)

Tragedy struck Alaska again on 22 November when another Air Force transport disappeared on a flight en route to Elmendorf AFB. Assigned to the Military Air Transport Service (MATS), the four-engine Douglas C-124A Globemaster

211

was carrying 52 passengers and crew from McChord AFB in Washington. It last reported over Middleton Island in the Gulf of Alaska.

Some of the military aircraft assigned to the missing C-119 search were reassigned to the subsequent search for the C-124A, leaving only ten aircraft involved in the original search operation. Even so, with improved weather conditions over the next few days the aircraft effectively covered a large portion of the C-119's estimated flight route. Additional snow slides observed around Mt. Iliamna and Mt. Redoubt were again investigated and pictures taken of the area, but nothing conclusive could be determined.

On 25 November wreckage from the missing C-124A was found on Mt. Gannett, thirty miles off course and only fifty miles east of Anchorage. There were

Four C-119 Boxcars preparing for a mass airdrop mission in Alaska, 1955. (USAF via A.T. Lloyd)

no survivors. The probable cause was later determined to be a navigational error, due to contributing factors of significantly higher than forecast winds and poor radio range reception caused by precipitation static.

Once the missing C-124A was located, more aircraft again joined the search for the lost C-119C. Efforts resumed on a daily basis from Elmendorf and Kodiak, as well as occasional planes taking part from Adak NAS in the Aleutians and Ladd Field in Fairbanks. Fifteen various military aircraft were still involved in the search through the first week of December. The entire route over land and water between Anchorage and Kodiak was eventually covered numerous times at various altitudes and airspeeds. Every potential piece of wreckage, possible crash site, reported sighting and unidentified signal was investigated.

By 15 December heavy precipitation during the previous month had blanketed much of the region in multiple layers of snow, making a realistic sighting of any wreckage virtually impossible. All search activities were officially terminated, with the hope future conditions might eventually reveal the location of the lost plane.

During the Air Force's official investigation of the missing C-119, every available piece of information was scrutinized as a possible factor in the accident. Although nothing conclusive could be determined, it was the Accident Board's opinion that the probable cause of the accident was pilot navigational error. Contributing factors were believed to be a weak radio signal from the Kenai Radio Range and difficulty in identifying the station because of precipitation static.

The Accident Board also concluded the co-pilot was not qualified for flight in instrument conditions and the aircraft was not adequately equipped for instrument navigation in the Alaskan Theater. All C-119C aircraft in the 534th Troop Carrier Wing at the time only had a single radio range receiver installed. It was determined the pilots could have easily asked for a radar fix from ground radar facilities in the area if they were unsure of their exact position, which had been briefed as an option to the crew before departure, but was not done.

C-119C over the Yukon, April 1951. (USAF via A.T. Lloyd)

Ironically, several reports submitted through the Air Force system prior to the accident detailed a need for two radio compasses in tactical aircraft flying in remote areas. Although C-119 aircraft did have the new, more accurate omni-range receivers on board, omni-range stations had not yet been made operational in Alaska.

After the findings of the Air Force Accident Report were released to the major commands for review, a controversy developed over a key issue. Several higher level commanders in the missing crew's chain of command thought the determination of a navigational error was wrong. This was based on three major points. One was the qualification of the crew, which included an experienced instrument instructor as pilot-in-command and the addition of a third pilot specifically assigned because of his familiarity with Alaska flight operations; both of which should have prevented any navigational errors. Second was the fact the unidentified radar plot could have been from a different aircraft entirely. Third and most important, C-119 aircraft in the Air Force inventory had a history of catastrophic propeller failures, which could have occurred in this case, causing a major structural failure and subsequent crash into the ocean.

C-119C aircraft were still relatively new arrivals to the Air Force fleet in late 1952. The first version of the Fairchild C-119 Flying Boxcar entered service in

A flight of C-119s on an airdrop mission over Korea. (USAF via Chuck Lunsford)

1949 as the C-119B, replacing the older C-82 Packets in use since 1945. Although similar in appearance to the C-82, C-119s had a wider and stronger fuselage and more efficient Pratt & Whitney engines for carrying heavier cargo loads over greater distances. With a maximum speed of 281 mph, range of 1,700 miles, a payload of 34,000 pounds and a ceiling just under 24,000 feet, the transports were capable airframes.

Regarded as a much needed improvement over the C-82, the C-119 still had its share of problems after entering production. The most significant was a tendency for propeller assemblies on the 3,500 hp Pratt and Whitney R-4360 engines to fail unexpectedly in flight, often causing a sudden and disastrous engine

Chapter 14—Operation Warm Wind

separation from the wing. Several previous accidents involving C-119 aircraft were attributed to propeller failure.

Only two days after the C-119C was reported missing in Alaska, another C-119 in Montana crashed as a direct result of in-flight propeller failure. A total of four C-119s were lost in the Air Force during a ten day period in November 1952 alone.

In February 1952 all C-119s in Korea were grounded after several unexplained accidents. The only one that could be explained was attributed to propeller failure. The Air Force command in Korea subsequently made several structural and propeller modifications to their C-119 fleet, which resulted in far fewer accidents. Unfortunately new Fairchild C-119 aircraft were still being produced at the factory without the new modifications and other Air Force commands failed to implement the same changes that had been completed in Korea. Aircraft 570, the missing C-119C, was one of those aircraft without the modifications.

The fact most of the crew was experienced and qualified did not by itself preclude them from making navigational errors. Similar fatal accidents in Alaska involving a C-119 on 7 November and a C-124 on 22 November, were attributed to navigational error, even though the two aircraft had highly experienced and qualified crews.

A study conducted by the Air Force concerning Air Force accidents between January 1947 and July 1951, revealed that out of 51 major aircraft accidents that occurred in that span, 45 were influenced by navigational errors. Six of the accidents involved large multi-engine aircraft flying into higher terrain. All of the navigational errors which occurred were caused by insufficient preflight planning, poor judgment or a lack of knowledge in the proper use of navigational facilities. Experienced pilots were involved in many of the accidents.

Following the series of fatal accidents in Alaska in 1952, an Air Force investigation of the radio navigational aids and communication facilities in the region found the overall effectiveness and safety of the airway systems to be lacking. Improvements were recommended to the responsible federal agencies.

In the five subsequent decades which have passed, twenty major military accidents involving multi-engine aircraft in Alaska have been attributed to navigational error. The mystery of the missing C-119C Flying Boxcar and its twenty occupants remains unsolved.

Chapter Fifteen

June 5, 1969

Last Flight of Irene 92

After lifting off from the hard surfaced runway at Shemya in the Western Aleutian Islands, the Air Force RC-135E climbed rapidly in a slow turn northeastward, leaving the remote, wind swept military base behind. Departure was at 7:02 in the morning and contact with the radar controller was established two minutes later. Upon reaching its assigned cruising altitude the highly secret aircraft notified the controller it was level at FL250 (25,000 feet) and proceeding on course direct to Eielson Air Force Base, approximately 1,700 miles to the northeast near Fairbanks, Alaska. Twenty miles from the field radar service was terminated by the Shemya controller and acknowledged by the crew.

The aircraft's flight path was intended to take it over the Bering Sea near the southern tip of Nunivak Island, then inland over the coast near Bethel and on into the interior of Alaska. A total time of four hours was estimated en route to Eielson AFB.

As Irene 92, code name "Rivet Amber," continued under the fresh morning sky with the sunlight reflecting off the surface of the fuselage, the nineteen occupants settled into their seats for the anticipated relaxing flight. Except for the crew in the cockpit, the other occupants were not even aware of the brilliant blue sky above or solid cloud layer below at 8,000 feet. Standard fuselage windows found on a civilian airliner were not present on the RC-135E, distracting from the specialized mission requirements of the crew. The purpose of the flight on this day was not to monitor secret Soviet military activity, as was usually the scenario, but to return to Eielson AFB for a routine maintenance inspection.

Forecast weather conditions along the route called for increasing variable cloud layers as high as 30,000 feet after takeoff, with in-flight visibility from one to three miles as the aircraft approached the mainland. Winds at flight level were predominately out of the south-southwest at 15 to 25 knots. No icing or turbulence of any kind was briefed for the flight.

"Rivet Amber" parked in a secure location at an unknown airbase. (King Hawes)

Aerial view of remote Shemya Island and airbase. (King Hawes)

Thirty-four minutes after lifting off from the runway at Shemya and sixteen minutes after reaching its cruising altitude, Irene 92 informed the Air Force controller at Elmendorf AFB near Anchorage they were experiencing in-flight vibrations, but were unsure of the malfunction and had the aircraft under control. The voice from Irene 92 stuttered slightly as if the person was experiencing some difficulty in breathing.

The Elmendorf controller responded, "You say you're not declaring an emergency. Is that Charlie (correct)?"

This was followed by noticeable, possibly distressed breathing from Irene 92 as a second radio message was attempted over a period of several seconds. "Roger... ahh... ahh... crew go to oxygen," which was transmitted at 7:38 am.

Elmendorf Control asked Irene 92 to repeat the transmission, but there was no vocal response, only the sound of the radio being keyed at different intervals. This continued for the next hour and six minutes, with Elmendorf Control unable to reestablish radio communications and the repeated keying of a radio transmitter from what was believed to be Irene 92. No further audible radio transmissions were received from the aircraft.

At a cruising speed of more than 430 knots, the last vocal transmission at 7:38 would have placed the RC-135E approximately 250 nautical miles northeast of Shemya Island and about 200 miles north of Amchitka Island in the Aleutians. During two previous radio transmissions with the Elmendorf controller, Irene 92 estimated crossing the 180 degree meridian at 7:43 local time. The 180 degree meridian was the next mandatory reporting point along the flight route.

Shortly after Irene 92 departed from Shemya it was no longer in radar contact. The ground controlled approach radar service at Shemya had limited range and coverage was terminated four minutes after takeoff. Radar service was not available again until crossing the 180 degree meridian, where the far limit of the Distant Early Warning Identification Zone (DEWIZ) began.

A communication search by civilian and military stations in the region was initiated for Irene 92 at 7:54 after it failed to report crossing the 180 degree meridian and no radar identification could be established. Once the plane's fuel reserve would have been exhausted at 2:55 pm, it was officially declared missing. Search efforts began even before Irene 92's estimated time of arrival at Eielson AFB had passed, since it was obvious the flight had experienced some sort of malfunction or emergency which precluded radio contact. The first aircraft to launch on the search was a KC-135 at 9:58 am, which was joined by a Coast Guard C-130 approximately an hour later. By evening numerous aircraft from various military bases in Alaska and the Pacific region had been launched or diverted from other missions and were actively engaged in a search of the area. Nothing was ever found.

Irene 92 was the only RC-135E ever built, and with a price tag of thirty-five million dollars was also the most expensive aircraft in the Air Force inventory at the time. In operation for less than two years, the RC-135E was a top secret and ex-

"Rivet Amber" a Eielson AFB; the only RC-135E model built. (354th Historical Office, Eielson AFB)

tremely specialized reconnaissance aircraft designed for the Cold War monitoring of military and nuclear weapons programs in the Soviet Union. Easily recognizable from the Boeing 707 civilian version, the RC-135E's exterior structure had been altered to house a powerful airborne radar system and the interior was completely modified with advanced electronic and surveillance equipment. Because of the sheer size and weight of the monitoring equipment being carried aboard, increased structural supports were added in the fuselage bulk heads and floors.

The 7- megawatt airborne radar array on the RC-135E was so powerful it re-

RC-135E "Rivet Amber" on a test flight over Alaska. The generator pod is visible below the left wing and heat exchanger pod is visible below the right wing. (E-systems-Big Safari via Ed Steffen and King Hawes)

quired its own power source, which was provided by an engine driven generator mounted in a pod under the inboard left wing. The massive amount of heat it generated required a separate heat exchanger under the opposite inboard wing. It was the most advanced airborne radar system at the time, able to accurately track a basketball sized object at a distance of more than 300 nautical miles.

Although the RC-135E was one of a kind, it supplemented many other RC-

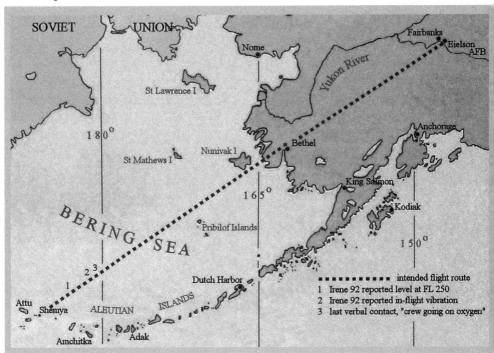

Intended flight route of Irene 92, code named "Rivet Amber," from Shemya to Eielson AFB, near Fairbanks, Alaska.

135s coming into service during the late 1960s. Many are still in service today. Variants of the Air Force RC-135 include various configurations for electronic and photographic surveillance that are an important part of the United States government's intelligence gathering capability. Just as they were in 1969, the aircrews assigned to the aircraft are hand picked, highly trained, expertly qualified and extremely motivated.

When Irene 92 first began operating from Shemya to monitor Soviet activities on the Kamchatka Peninsula in 1968, she was supplemented by an RC-135S, code named "Rivet Ball", configured primarily for photographic intelligence. Both were assigned to the 6th Strategic Reconnaissance Wing at Eielson AFB. Although both RC-135s routinely operated near the Soviet border in international airspace, neither was configured with any armament and if attacked by hostile aircraft would have been completely defenseless. They operated as a team until January 1969 when "Rivet Ball" was heavily damaged during a crash landing at Shemya. Beyond repair and gutted of its top secret surveillance equipment and useful components, what remained of the aircraft was later destroyed. Two additional RC-135S aircraft, known as "Cobra Ball I" and "Cobra Ball II," replaced the two previous aircraft in late 1969 and early 1970.

In the hour preceding Irene 92's departure, a standard mission briefing was conducted by the crew for the routine flight to Eielson AFB. Weather conditions were favorable and maintenance on the aircraft was always kept at a high level. A pre-flight inspection had been performed the previous day. The aircraft was kept in the hangar over night and was only pushed outside when the crew arrived the following morning. Engine startups, taxi and takeoff procedures were normal, with the crew giving no indications of any problems.

One of the replacement Cobra Ball RC-135Ss being refueled by a KC-135 tanker. (354th Historical Office, Eielson AFB)

Within a few hours after Irene 92 disappeared aircraft and surface vessels reached the area of its last known position and began a systematic search. The long hours of summer daylight provided visible flight conditions well into late evening, and after nightfall several aircraft continued searching using aerial flares and electronic monitoring equipment. Two Coast Guard cutters which arrived on scene continued a sweep of the waters looking for possible survivors or debris. Nothing was found.

During the first night of the search a report was received from one of the search aircraft of a possible radio transmission on the emergency guard frequency, but it was erratic and could not be understood or the location identified. The signal was similar to what would be sent by an emergency rescue radio carried aboard the missing RC-135E.

By the second day of the search operation fifteen military aircraft were involved from bases at Kodiak, Adak, Elmendorf, Eielson, Hawaii, California, Japan and Okinawa. Under the jurisdiction of the 17th Coast Guard District in Juneau, aircraft and surface vessels were coordinated to search in a grid pattern along Irene 92's proposed flight route, focusing primarily on the area of its last confirmed transmission. Winds were reported as 20 knots on the surface with six foot seas and ten miles visibility.

"Rivet Amber" at Eielson AFB. The heat exchanger pod is clearly visible under the right wing, inboard to the fuselage. (354th Historical Office, Eielson AFB)

Irene 92 carried two twenty-man life rafts as well as individual water survival equipment, which was necessary for any prolonged exposure in the cold waters of the Bering Sea. If the plane had gone down in the ocean and the crew transferred safely into the life rafts, the chance of a prolonged survival was excellent. But if any of the crew were without proper survival equipment and immersed in the cold 38 degree water for more than a few minutes, they would have quickly succumbed to hypothermia.

Each assigned grid in the search area was covered by aircraft flying a few hundred feet above the water and as slow as possible to effectively scan every foot of the surface. Primarily looking for oil slicks, debris or orange markings associated with individual flotation devices and rafts, the crews also watched for signals such as flares or smoke from possible survivors. If something was sighted that could not be positively identified from the air, a surface ship would be called into the area to investigate further.

Floating seaweed, kelp and garbage were not uncommon and at times were mistaken as a positive sighting of aircraft debris. This distracted from the overall search effort, but also eliminated any possibility something had been overlooked. Every potential piece of evidence was investigated, but by the end of the day nothing conclusive was found.

The second evening the weather deteriorated over the Bering Sea, bringing high winds and low ceilings that interfered with search aircraft operating through the night. Eight foot white-capped seas also developed on the surface, hiding potential clues in the rolling waves and ocean froth that limited visibility.

On 7 June more planes and ships joined the search, including several Japa-

RC-135E "Rivet Amber" lifting off the runway at Offutt AFB, Nebraska, before deployment to Alaska. (King Hawes)

nese and Russian fishing trawlers in the area. A mistaken report also materialized from a Japanese vessel when the captain confused a request in locating survivors with an actual report of survivors. Another message from one of the search aircraft reported the possible sighting of a life raft, but it later proved to be only an orange buoy marker that had broken away from a commercial crab pot. Several other aircraft saw possible parachutes or wreckage in the water, but all turned out to be false sightings after surface vessels were directed into the area.

Twelve to fifteen aircraft remained involved in the search over the next several days with nothing conclusive being found. Regular reports of unidentified debris continued to provide hope some evidence of the aircraft or its crew might be found, but after each sighting was investigated and dismissed as a false alarm, hope slowly faded.

Eventually the search expanded along the entire flight route from Shemya to the mainland, only to intensify again in the area of the plane's last transmission after additional reports of possible debris were once again sighted. Those too proved without merit.

On 13 June the Coast Guard called off their portion of the search after nine days of extensive effort, returning the Coast Guard aircraft and ships to normal duties. The Air Force persisted with eighteen aircraft for another full day to take advantage of good flying weather, before terminating the search completely. Nothing was ever found of the aircraft or crew, not even an oil slick which might have marked where the aircraft went down.

In all, the search aircraft and vessels logged several thousand hours in the air

Hangar #3 at Shemya AFB, the home of "Rivet Amber." (Logan Delp via King Hawes)

and on the sea. The various search aircraft alone logged 300 sorties and more than 2,500 flight hours, covering more than 435,000 square miles over the extent of the search. Figures released by the USAF estimated there was at least an 80% probability of detection over the entire search area and a 99% probability of detection in the most likely area covering 50,000 square miles around the missing aircraft's last radio transmission.

Speculation arose after the search began that the RC-135E probably experienced a catastrophic structural failure soon after making its last radio call. Because there was no definite position the aircraft had crashed and because of the sheer expanse of ocean to search, the likelihood of finding any wreckage was limited from the start. There was also the distinct possibility the aircraft had disintegrated in-flight, scattering small pieces of debris across a wide area, making it much harder to find.

A sonographic-audio analysis of the Elmendorf Control tapes by the FBI laboratory found some interesting information that disputed earlier speculation. The laboratory concluded dozens of transmissions from Irene 92 were attempted more than an hour after the aircraft first experienced in-flight difficulties. This analysis was based on harmonics and tones present in static transmissions over that period which were identical to harmonics and tones in Irene 92's previous transmissions. Included in the background noise was the extremely faint sound of a RC-135 warning horn, indecipherable speech syllables and deep breathing, all mixed in with static that was consistent with Irene 92's radio transmitter. Other tones matched those compatible with a malfunctioning power supply on the RC-135, and a faint voice attempting contact with the Elmendorf controller. Some audio portions of the tape also included apparent Morse code signals, but they were distorted and could not be deciphered. Transmissions identified as

A side view of RC-135E "Rivet Amber" in flight. (King Hawes)

originating from Irene 92 continued intermittently until 8:42, an hour and four minutes after the emergency first occurred, including faint voice sounds, Morse code signals and the blank keying of a transmitter. Additional strange tones from a transmitter that could not be confirmed as coming from Irene 92 continued until 9:27. Nothing further was recorded which could be construed as coming from the missing RC-135E.

The possibility the plane attempted to reach one of the islands to the south of its position or return to Shemya during the sixty-four minutes after experiencing a mechanical failure would have been the most likely scenario. It is also possible it continued along the proposed flight route toward the mainland, only to crash en route and much further away from the area of its last communication. Since the aircraft was obviously experiencing radio and power supply problems, difficulty with the navigational equipment would also have been occurring and could easily have misdirected the crew to fly off course from their intended destination. Lower cloud layers present over the Bering Sea and Aleutian Islands would have masked any visual sightings of the surface.

Even at a reduced airspeed after the initial in-flight difficulty, Irene 92 could have flown north or west out of the established search area over the next hour and four minutes. At the time of the extensive search activities the audio tapes had not yet been analyzed, and it was assumed Irene 92 probably went down near its last known position short of the 180 degree meridian. If false, that assumption would explain why not a single piece of wreckage or debris was found, even though both the primary and secondary search sectors were systematically and thoroughly covered.

Civilian "Tech Reps" who helped maintain the aircraft systems, tracking radar and navigation equipment on "Rivet Amber." (Logan Delp via King Hawes)

Another theory put forth by some individuals speculates the top secret aircraft was shot down by a Soviet spy ship waiting for an opportunity in the Bering Sea near Shemya. Certainly it is possible, but highly unlikely, since a deliberate pre-planned attack could have easily sparked Word War Three. If such an outcome were possible, however, it would certainly have been more feasible to shoot down the aircraft while it was flying an operational mission near the Soviet border, and not while operating so close to Alaskan territory where the likelihood of finding debris or survivors would have been much higher.

Soviet fishing vessels were operating in the area at the time of the disappearance, and throughout the Cold War some of those vessels were specifically configured as electronic spy ships, some possibly armed to prevent boarding or capture. The Soviet government was aware of the RC-135's mission and capability, and would have done almost anything short of war to shoot one down or capture the crew. Perhaps too, Irene 92 was not on a routine flight home as claimed, but on another top secret surveillance mission of Soviet activities that turned tragically hostile.

During the Cold War era dozens of American and Allied aircraft were fired upon or shot down by Soviet aircraft, while others disappeared while operating near Soviet airspace. There are also instances of American airmen reportedly being captured, but never released.

Whatever happened to Irene 92 will remain a mystery until some positive evidence can be found to explain the disappearance. A document might yet be uncovered deep in the Pentagon or old Soviet archives that could help reveal the

Last known image of "Rivet Amber" taken with an 8mm home movie during departure from Shemya in 1969. (Lt. Col Bob Brown, USAF Retired, via King Hawes)

plane's true fate. Then again, perhaps it is as simple as a fishing vessel some day pulling up pieces of wreckage from the bottom of the Bering Sea, or someone finding a half buried piece of debris on a remote shoreline where a storm deposited it above the tide-line decades earlier.

What is known is the nineteen crewmembers of Irene 92 lost their lives in the service of their country on that fateful day. Their sacrifice should never be forgotten.

In memory of Major Charles Michaud, Major Peter Carpenter, Major Richard Mortel, Major Rudolf Meissner Jr., Major Horace Beasley, Captain Michael Mills, Captain James Roy, Master Sergeant Herbert Gregory, Staff Sergeant Lester Shatz, Staff Sergeant Richard Steen Jr., Staff Sergeant Roy Lindsey, Staff Sergeant Robert Fox, , Tech Sergeant Donald Wonders, Tech Sergeant Charles Dreber, Tech Sergeant Eugene Benevides, Tech Sergeant Harvey Hebert, Sergeant Douglas Arcano, Sergeant Sherman Consolver Jr., and Sergeant Lucian Rominiecki.

Chapter Sixteen

October 16, 1972

Accident or Conspiracy? The Disappearance of Congressmen Boggs and Begich

Pulling himself up on the cabin deck of the Bell helicopter, a young mechanic slipped and lost his footing while reaching to remove the engine exhaust cover. Only his hand grabbing the cowling as he slid sideways saved him from falling completely off the deck onto the hard concrete several feet below. Pedaling his feet while dangling with one arm on the helicopter, he managed to catch the skid with the toe of his boot and regain an unbalanced foothold before lowering himself back down to the ground. Cursing his own stupidity for being in a hurry and not wearing gloves, he looked around the ramp to see if anyone had witnessed his near accident, then walked carefully back to the nearby hangar, flexing the sore fingers of his hand caught by the rough edge of the cowling.

A thin layer of ice still coated the surface of the helicopter. Earlier that morning, falling rain from a warm, upper air mass had passed through colder air near the ground and froze on contact with the metal of the aircraft. All around the airfield the super-cooled water droplets had adhered to anything left outside, leaving a dull, rough finish of ice that slowly began to melt during the increasing temperature of the morning sunrise.

By the time the helicopter mechanic returned outside with a folding stepladder, a Cessna C-310 airplane from the adjacent hangar was being wheeled onto the parking ramp. He watched as it was pushed clear of the large sliding door and moved into position in front of the nearby refueling site. The mechanic recognized Don Jonz, the owner and chief pilot for Pan Alaska Airways, who supervised the refueling before climbing inside the cockpit. As a light drizzle again began to fall, the mechanic glanced skyward into the morning overcast, wondering where the twin-engine Cessna was heading, grateful his own aircraft was on a weather hold until the forecast improved.

Diverting his attention back to the helicopter, the mechanic paid little further attention as the Cessna 310's engines started a few minutes later, followed by the aircraft powering down the taxiway and parking near the control tower, as if waiting on passengers. The mechanic did not notice the two important looking individuals and another carrying a small amount of baggage approach the plane, but several witnesses nearby recognized United States Representatives Hale Boggs from Louisiana and Nick Begich from Alaska, accompanied by Begich's special assistant Russel Brown. The three passengers and baggage were loaded aboard by Don Jonz, who then restarted the engines and repositioned for departure.

At 8:59 am the white and orange-striped plane lifted off from runway 24 and turned on a downwind departure southeast toward Turnagain Arm. It was last observed climbing through 2,000 feet by the tower controller.

Weather conditions at Anchorage were better than other airports in the area and especially the surrounding mountain passes, which precluded visual flight due to low ceilings and visibility. Jonz had received a weather briefing earlier in the morning for Juneau, Sitka, Yakutat and Cordova, detailing the poor conditions expected on a proposed route from Anchorage to Juneau. In the Cook Inlet area a significant weather advisory was issued for moderate to severe turbulence with strong gusting winds and rain showers. The area forecast for Cook Inlet predicted moderate rime icing in clouds above 6,000 feet, with Portage Pass closed due to low ceilings and visibility. Portage Pass was a well used visual flight route through the mountains on the eastern end of Turnagain Arm. Forecasts for the North Gulf Coast and Southeast Alaska, as well as terminal forecasts for Cordova, Yakutat and Sitka, all called for continued ceilings as low as 500 feet and visibility as low as a half mile in fog. Visual flight was not recommended.

At 8:40 am a pilot on a U.S. Air Force helicopter flying to Seward reported experiencing moderate to severe turbulence on the west side of Portage Pass, headwinds of fifty-five knots and a low overcast with reduced visibility further east. In-flight conditions forced him to alter his route south along the highway in order to stay clear of the pass.

In spite of the weather forecast and pilot report, Jonz continued flying under visual flight rules (VFR) from Anchorage to Juneau. Ten minutes after departure, approximately half-way to Portage Pass, he filed a VFR flight plan with the Anchorage Flight Service Station (FSS), giving a proposed route along the V-317 airway to Yakutat, then direct to Juneau. Estimated time en route was three hours and thirty minutes at 170 knots airspeed. Six hours of fuel and four people were stated to be on board, as was emergency gear and a required emergency locator beacon, in response to a direct question from the FSS controller. The current route and destination weather conditions were also reiterated to Jonz by the station controller. It was the last contact with the flight and no evidence of the aircraft or occupants has since been found.

Within an hour after the flight's failure to arrive in Juneau, the most massive search in United States history ensued for the lost aircraft carrying two U.S. Congressmen. Influenced in large part by political pressure from Washington, D.C., the extensive search operation would last five and a half weeks, involving numerous government and civilian agencies and hundreds of individuals. Every means available was utilized to find the missing plane, from advanced optical

A parked C-310 displaying the streamlined silhouette that made it popular in the civilian market. (1000aircraftphotos.com)

and infrared photography to electronic monitoring, aerial reconnaissance and ground searches, without a single clue revealed. The probable cause of the disappearance could not be determined. Many claim it was an unfortunate accident, while others claim it was a deliberate act of sabotage involving the highest levels of government.

As the U.S. House of Representatives Majority Leader, Hale Boggs was the second highest-ranking Democrat behind the Speaker of the House, and was regarded as the likely choice to ascend to that position once the acting speaker retired in 1978. He was one of the most influential members of Congress, having served twenty-eight years as a Representative from the State of Louisiana, including a key position on the House Ways and Means Committee and as a member of the controversial Warren Commission which investigated the assassination of President John F. Kennedy.

Congressman Hale Boggs was in Alaska on a fund-raising trip for his fellow

Democrat and protégé Nick Begich, who was campaigning for only his second term as the U.S. Representative from Alaska. They attended an expensive cocktail party and banquet for local Democratic Party faithful in Anchorage the night before they disappeared, raising more than $20,000. On their way to Juneau the next morning for another campaign dinner, Boggs was the scheduled speaker at a posh event that evening, planning to continue onto Washington later that night.

Warren Commission members present their findings on President Kennedy's assassination to President Lyndon Johnson, September 1964. Congressman Boggs is furthest on the right. (LBJ Library and Museum-Austin, Texas)

Nick Begich was elected as Alaska's Representative to the U.S. House in 1970. As a first term Congressman he served on the Public Relations and House Interior Committees, and was instrumental in settling Alaska Native land claims necessary for construction of the Trans-Alaska pipeline.

An Air Force HC-130 from Elmendorf Air Force Base (AFB) outside Anchorage, already airborne on a separate mission, was the first aircraft to begin searching for the missing Cessna 310C the afternoon it disappeared. Assets from the Coast Guard and Alaska Civil Air Patrol in Anchorage, Yakutat and Juneau were involved a short time later. Visual and electronic sweeps of the route from Anchorage to Juneau were also conducted, although Portage Pass and other coastal areas remained obscured by low clouds.

HC-130s remained airborne the first night and continuously for the next twelve days monitoring possible electronic signals from the plane's Emergency Locator Transmitter (ELT). After twelve days the monitoring was scaled back and only conducted when other search aircraft were actively involved. Several days into the search it was revealed the missing plane probably did not have an ELT aboard, as claimed by the pilot when he filed his flight plan. The ELT personally owned and carried by Jonz was found in another company aircraft, and employees who were inside the plane shortly before takeoff stated they saw no evidence of one onboard.

ELTs were designed to automatically transmit emergency distress signals during a crash that could be monitored by any of several search and rescue satellites or other aircraft. The signal could then be traced to a general area or more specific location depending on the type of equipment receiving the signal. A battery on an ELT operating in automatic mode could transmit for forty-eight continuous hours, or for weeks if turned on and off manually for only short periods of time. All aircraft operating in Alaska were required to have one aboard.

No survival equipment was observed in the aircraft by employees before the flight either, which

Representative Nick Begich on his first term in office. (Library of Congress)

was also a requirement for flight operations in Alaska. Pan Alaska Airways did own three containers of survival equipment, but all three were located in the company's main Fairbanks facility after the plane disappeared.

The twin-engine Cessna C-310C flown by Jonz and carrying U.S. Congressmen Boggs and Begich and his personal assistant Russel Brown, was a dependable aircraft used by commercial air charter operators and private pilots. It was equipped for instrument flight and although it had an oxygen system installed for flight above 10,000 feet, the system had not been serviced with oxygen. There was also no auto-pilot system or anti-icing capability on the aircraft, which was probably the reason Jonz elected to fly on a VFR flight plan.

The Cessna Aircraft Company first launched the production model Cessna 310 in 1954 to meet the growing demand in civil aviation for a small, fast, comfortable, multi-engine plane capable of carrying several passengers. Certified for instrument flight and equipped with retractable landing gear, a steerable nose wheel, constant speed feathering propellers and two wing tip fuel tanks, the aircraft became an instant hit. Cessna A and B models were virtually identical and reached market in 1957. Both had an increased gross weight and optional auxiliary fuel tanks. The Cessna A model was a military version given the designation L-27A or U-3A, while the B model was its civilian counterpart. C model Cessna 310s were introduced two years later with higher horsepower, fuel-injected engines, increased gross weight, a maximum speed of 242 mph and a ceiling of 21,300 feet.

Side view of a Cessna 310 in flight. (1000aircraftphotos,com)

During the first four days of the search, at least thirty military, Coast Guard and Civil Air Patrol aircraft were actively taking part, limited only by bad weather and available daylight. Coast Guard vessels from Kodiak to Ketchikan were also dispatched to help cover the search areas by sea, and military ground teams were organized to begin searching on foot as they were needed.

Three general areas became the primary focus of the search operation, with each area broken down into smaller sectors and assigned to individual aircraft, surface ships and ground teams. The primary search area included approximately 34,000 square miles along a tapered corridor which extended outward for six hundred miles from a ten mile wide circle around Anchorage to a fifty mile wide circle around Juneau. A secondary search area covered 53,000 square miles on either side of the primary area, running along a wider corridor from a twenty mile circle around Anchorage to a seventy mile circle around Juneau. A third search area en-

Chapter 16—Accident or Conspiracy? Disappearance of Congressmen Boggs and Begich

compassed possible routes the missing aircraft might have flown between Anchorage and Sitka, which was often used by pilots as an alternate destination in inclement weather. Weather conditions at Sitka were specifically requested by Jonz and briefed by the FSS controller the morning his plane disappeared.

As the search continued with little evidence as to the fate of the missing Congressmen and their plane, the House of Representatives in Washington, D.C. began conducting daily briefings on the status of their colleagues. The search soon intensified as more and more government assets became available, even classi-

USAF Lockheed C-130 search aircraft used in the search for the missing Congressmen. (USAF via Howard Wynia)

fied military hardware. A top secret SR-71 spy plane was dispatched to Alaska and photographed the entire search area using infrared technology, providing thousands of high resolution images, and two Air Force RF-4C all weather reconnaissance planes began sweeps of the area using advanced optical and electronic technology. Dozens of technical and intelligence experts analyzed every piece of collected data, but nothing significant was found.

By the fifth day of the operation more than forty aircraft and four Coast Guard cutters were involved, and those were supplemented by numerous fishing vessels and commercial planes transitioning through the same areas. Storm-tossed seas made it extremely difficult in spotting possible floating debris, and the tides and currents would have beached or carried further out to sea what little remained after a few days. Even if the plane crashed in the water and the occupants survived the initial impact, there exited a strong probability they would have succumbed to the cold water in a matter of minutes. Without a flotation device it was unlikely any bodies would even have remained afloat to be discovered.

Because the last contact with the plane was made before entering Portage Pass and the weather in the pass was unsuitable for visual flight conditions, it was believed by many of the pilots involved in the search that the missing Cessna 310C most likely crashed in that area. Pilots considered the pass dangerous, even on clear days when high winds funneled through the mountains, often causing violent turbulence and strong downdrafts. On days when high winds were combined with poor visibility, low ceilings and icing, visual flight through the narrow pass was a disaster waiting to happen. The same day Jonz and his passengers disappeared, another plane attempting to fly through the pass turned around due to severe turbulence, and a third was forced to climb above the clouds after encountering severe icing. The third aircraft was also a C-310, but unlike Jonz's plane was equipped with supplemental oxygen, allowing it to fly above 10,000 feet.

SR-71 high altitude reconnaissance aircraft used in the search for Congressmen Boggs and Begich. (USAF via Howard Wynia)

Portage Pass is the main transition point for aircraft flying through the coastal Chugach Mountains between Turnagain Arm and Prince William Sound. The lowest elevation of the pass is at 400 feet, with steep snow-covered peaks within a few miles north and south rising several thousand feet higher. Notorious for its high winds, the pass has claimed many small aircraft in the decades before and since. Other aircraft besides Jonz's Cessna 310 have disappeared in the same area over the years, likely hidden beneath an avalanche of snow on the side of a rugged slope or in one of the many glacial ravines where they were violently forced earthward by unpredictable winds.

Chapter 16—Accident or Conspiracy? Disappearance of Congressmen Boggs and Begich

Unconfirmed reports from other areas along the Anchorage to Juneau route, however, indicate the missing aircraft might have made it through the pass and continued toward its destination. One person in Whittier, near the east side of Portage Pass, came forward claiming to have heard a plane in the area around 9:30 the same morning Jonz and the Congressmen disappeared, which was the approximate time they should have been in the area after flying through the pass.

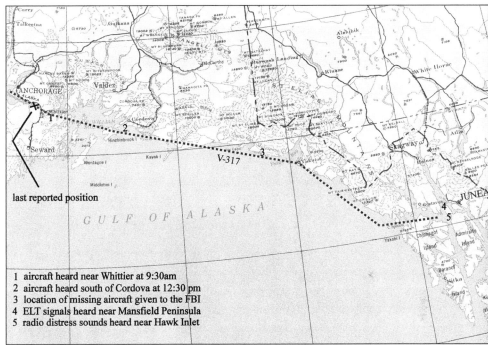

Map of Alaska showing the locations where the missing C-310C was reported the day it disappeared.

Another believable report came from hunters twenty miles south of Cordova, who heard a twin-engine aircraft circling above the overcast for several minutes before heading out over the water. The time was around 12:30 on 16 October, only three and a half hours after Jonz departed from Anchorage.

The day after the Cessna vanished, three partial ELT signals were monitored near the Mansfield Peninsula, fifteen miles southwest of Juneau, but a Coast Guard helicopter dispatched to the area did not locate any evidence of the airplane. The following day several individuals from a logging camp further southwest of Juneau and a local pilot flying in the area, reported hearing strange sounds the afternoon the C-310C disappeared, as if an aircraft was transmitting a distress call. Three Coast Guard helicopters searched the area extensively, without success.

Attention stayed focused around Juneau for several more days after strange Ham radio signals from an unidentified aircraft were reported to have been monitored by five different individuals the day the missing plane disappeared. A series of conflicting transmissions from the pilot, monitored by the Ham radio operators, stated he was encountering strong headwinds and was low on fuel. The Ham radio operators heard the pilot say he had fifteen minutes of fuel remaining and was a hundred miles over water, yet two hours later a second message from the pilot stated they were about to crash on a rocky beach. A third transmission from the pilot claimed there were three injured passengers and they were down twelve miles southwest of Juneau. At no time did the pilot identify himself or the aircraft call sign.

What seemed incredulous was the fact the radio transmissions were monitored five hours after the C-310's fuel would have been exhausted and the Ham operators who heard the distress calls all resided in California. The time discrepancy was explained in one of the messages from the pilot who indicated he had landed on a remote airstrip earlier in the flight to wait for better weather conditions. Why the radio transmissions were only heard in California was not explained, but it is not unheard of for radio signals to bounce erratically off the atmosphere under certain conditions, only to be monitored thousands of miles away. The reports were taken seriously and thoroughly investigated, but nothing conclusive was determined.

The coordinated search operation was not without its reports of less believable information as well. Several individuals with supposedly psychic connections came forward offering their assistance, but usually with a price. One self proclaimed mystic flew to Alaska at his own expense and identified what he believed was the location of the missing plane near Snowshoe Lake, located off the Glenn Highway between Anchorage and Glennallen. The site was well north of the plane's proposed route over Prince William Sound, but was still investigated by a search aircraft. Nothing was found.

One of the most controversial claims involving the disappearance of the two Congressmen occurred shortly after the search began. Evidence was presented by a mysterious person possessing sensitive information, who plotted the exact location of the aircraft six miles inland from the coast, on a glacier northwest of Yakutat. Statements received by the FBI and a further investigation of the individual and company involved in testing reportedly technologically advanced surveillance equipment, were never confirmed or repudiated and partial copies of the FBI documents were not released until 1992 under the Freedom of Information Act. The individual who first came forward with the information claimed a company testing a sophisticated electronic satellite surveillance system had accurately tracked the plane from Anchorage until it crashed on the glacier, and was further able to track two apparent survivors walking away from the plane a short time later.

The information was relayed to the Coast Guard's Eleventh District Chief of Operations. Although there is no indication a specific search of the area was initiated based on the information provided, the indicated crash site was within the primary search area covered on numerous occasions during the operation. The seemingly detailed facts furnished by the unnamed person and lack of supporting evidence of a further investigation into the claims, has led to subsequent accusations of a government cover up and conspiracy from several sources.

Other claims from the same sources link Congressman Hale Boggs and his involvement in the Warren Commission as a motive for the conspiracy, based on

Congressional leaders speaking to the press, July 1964. Congressman Boggs is fourth from the left, behind the two at the podium. (LBJ Library and Museum-Austin, Texas)

the supposed intent of the Congressman to release newly obtained information on the assassination of President Kennedy. The information presumably linked Richard Nixon to the assassination and was the motive for the Congressman's silence. Hale Boggs had also publicly accused the Director of the FBI, J. Edgar Hoover, of illegal wire-tapping and reportedly demanded a new investigation into the assassination of President Kennedy only a week before he disappeared.

Whether theories of sabotage and conspiracy in the disappearance of the two U.S. Congressmen are actually true is unknown, but accusations persist today among various individuals and organizations. A witness who was with Jonz before the flight said he carried a small, unidentified metal object aboard in a briefcase. It was the approximate size of an ELT, but not the same color of any owned by Pan Alaska Airways. Perhaps it was some perfectly innocent and unrelated device, but for anyone giving credence to charges of sabotage, a more sinister function cannot be dismissed.

One interesting bit of information regarding the disappearance of Hale Boggs mentioned a young up-and-coming democrat, who was involved in the presidential campaign for George McGovern at the time and drove Hale Boggs to the air-

Congressman Hale Boggs meets with President Lyndon Johnson in the Whitehouse, May 1968. (LBJ Library and Museum-Austin, Texas)

port on his fund-raising trip to Alaska. The man was William Jefferson Clinton, the future president of the United States, whose own controversial political career would be surrounded by allegations of murder, corruption and misuse of power.

Weather conditions that existed the day of the flight and Jonz's own history reveal a more probable cause of the accident. Perhaps Don Jonz felt pressured to fly the Congressmen in spite of the poor weather, or became overconfident in his own abilities. Hindsight can often be used as a crutch when writing about an event after the fact, but with more than 16,000 flight hours to his credit, Jonz certainly should have known better than to continue VFR in instrument conditions, especially without anti-icing or supplemental oxygen available on the aircraft.

Severe turbulence and icing did exist in the area, combined with low ceilings and visibility. Either condition could have caused a catastrophic structural failure with little warning.

In 1966 Jonz had his Air Transport Pilot license suspended by the FAA in Miami for violating several flight regulations. It was reissued in 1968 and at the time the FAA acknowledged he was completely qualified and certified as a pilot. He also crashed a plane a few years before the disappearance in the mountains near Coldfoot, Alaska, but was not seriously injured.

Don Jonz had been flying in Alaska for more than fifteen years, but was reportedly not experienced with coastal conditions peculiar to Southeast Alaska that were far more unpredictable. He was well regarded by some pilots who knew him, but also considered a "risk taker" and arrogant about flying in poor weather by others. By simple fate or coincidence, an article on flying in icing conditions he penned for *Flying* magazine shortly before his disappearance, was found informative but also cavalier to some pilots for not portraying icing with the respect it deserved.[1]

The article in *Flying* and another in a *National Pilots Association Service Bulletin*, however, also addressed significant factors on flying in hazardous weather conditions which showed an acute awareness of flight preparation and decision making.[2] Jonz specifically advised in his articles against operating aircraft in bad weather without maintaining a mental picture of the weather ahead, and the consideration for changing altitude, carrying extra fuel and always having an alternate course of action. He even made a point of informing pilots to ensure the proper survival equipment was aboard before each flight. For some reason Jonz disregarded his own advice on the day he disappeared.

The most intensive search operation in American history was finally terminated on 24 November after thirty-nine days of extensive coverage. Every potential lead and sighting was thoroughly investigated, no matter how unbelievable or inconsequential it seemed. Almost a hundred sightings and reports were received during the five week operation, none of which provided any positive clues leading to the discovery of the missing plane. More than 3,600 hours were flown by military and Civil Air Patrol aircraft. Hundreds more hours were spent on the water by surface ships and Army ground teams covering mountainous terrain.

In all, 325,755 square miles of remote and inhospitable territory along the Alaska coast was searched during the operation, many areas on multiple occasions. More than half of the total area was photographed with infrared and optical equipment by military reconnaissance aircraft, while electronic and observa-

---

1 "Ice Without Fear," *Flying* magazine, November 1972.
2 "Light Planes and Low Temperature," National Pilots Association Service Bulletin, Vol. XII, No. 1, January 1972, and Vol. XII, No. 2, February 1972.

tion aircraft effectively covered the entire region. The most likely areas where the plane might have crashed were covered a minimum of thirteen times and some as many as seventeen times. In spite of an estimated 99% probability of locating the missing plane, as determined by the military, no evidence was found of the aircraft or its four occupants.

The chance of nothing being found, considering the volume of assets and technology involved, seems highly improbable. Unlike other massive searches in Alaska and other parts of the world, the search for the missing Congressmen had every advantage; advanced electronic and surveillance equipment, access to sensitive information, the participation of government intelligence agencies and almost unlimited transportation resources. So why did the search fail? Were other influences working behind the scene to divert the investigation, as some claim, or was a catastrophic explosion or extreme turbulence at fault, leaving nothing of consequence to be discovered?

Then again, maybe the estimate of success was a gross exaggeration and there was never much chance of finding the plane in such an enormous and remote geographical area. It might yet reside on a melting glacier or tree shrouded island, intact and holding secrets, waiting to be found.

Chapter Seventeen

February 10, 1977

Cleared As Filed

For most Army pilots assigned to the 222nd Aviation Battalion's Command Flight Platoon the mission would have been considered another routine passenger flight. Based at Elmendorf Air Force Base in Alaska, the platoon pilots were often flying to one of many military installations scattered across the State. Long hours were the norm, with little time for a decent meal or relaxing cup of coffee. This day would be no exception.

Only Captain Deeter, the co-pilot on the mission, seemed enthusiastic about the trip due to his recent assignment in Alaska and his first time flying between Elmendorf AFB and King Salmon. He was assigned on the mission so his required orientation and instrument training could be completed at the same time. The pilot-in-command (PIC) was Chief Warrant Officer Four (CW4) Battle, an experienced instructor pilot and instrument examiner who had been stationed in Alaska for several years.

Aircraft in the Command Flight Platoon were used primarily for command and staff missions in direct support of the 172nd Infantry Brigade, headquartered at Fort Richardson, next to Elmendorf Air Force Base outside of Anchorage. Aviation support for other units within the Alaska Command was provided on a lesser scale, as availability of aircraft and scheduling permitted. Missions included regularly scheduled flights to Fort Wainwright and Fort Greely several times a week, training flights and occasional support missions to other bases throughout Alaska and the continental United States. Considering the limited number of personnel and aircraft in the platoon, the unit was one of the busiest flight sections in the Army.

Assigned aircraft in the Command Flight Platoon included the twin-engine

Beech U-21As, U-21Fs and C-12s, which were military designations for the civilian A90, A100 and A200 King Air, respectively. All three were proven and dependable airframes in both the military and civilian aviation communities. Versions of the U-21 had been in military service since 1967 and the C-12 since 1973.

U-21As were equipped with 550 hp Pratt and Whitney PT6A-20 turbo-prop engines, providing a maximum airspeed of 265 mph and a range of 1,675 miles. They could carry a maximum gross weight of 4,220 pounds, including crew, passengers, fuel and baggage. Seating was provided for two pilots in the cockpit and eight occupants in the cabin.

Lithograph print of a U-21A in flight, copyright by aviation artist Paul Fretts. (Paul Fretts)

For the roundtrip instrument flight from Elmendorf AFB to King Salmon, the U-21A was scheduled to carry the incoming Commander of the 1/43rd Air Defense Artillery Battalion, Lieutenant Colonel John Edge, as well as the outgoing Commander, Lieutenant Colonel William Barrett, the Operations Officer, Major James Nelson, and the Battalion's senior enlisted soldier, Command Sergeant Major Melvin Swiney. The crew consisted of the two pilots, CW4 Ralph Battle and Captain Donald Deeter, and a crew chief, Specialist Paul Jones.

Because of the scheduled early departure, CW4 Battle began the pre-mission

Chapter 17—Cleared As Filed

U-21 "Miss Piggy" of the 222nd Aviation Battalion being preheated at Barrow, Alaska. (Francis Boisseau)

planning the day prior. A route utilizing the V321 airway from the Homer VORTAC across Cook Inlet to King Salmon would normally have been used, based on the aircraft being unpressurized and the minimum en route altitude over the airway being below the level where supplemental oxygen was required.

U-21A performing an engine run-up check. (US Army Aviation Museum –Ref. Collection PN6021)

Although U-21As were equipped with supplemental oxygen in the form of individual face-mask type regulators, their use was considered an inconvenience to passengers in the cabin. Because of that, U-21A aircraft with passengers, especially high ranking passengers, were usually operated below an altitude requiring supplemental oxygen.

On the day of the scheduled mission, however, the TACAN portion of the

Homer VORTAC was out of service, causing some concern from the pilots about being able to accurately fix points along the V321 airway.[1] Instead, an alternate route using a different victor airway with a higher minimum en route altitude was selected.[2] That route was farther north of Homer and crossed Cook Inlet using the V456 airway from Kenai to King Salmon. Aircraft operating on the V456 airway were required to maintain a higher minimum en route altitude after passing TUCKS intersection, which changed from 5,000 to 13,000 feet. This increase in altitude necessitated the use of supplemental oxygen by the passengers and crew.

The morning of the flight the crew arrived early to finish their flight planning and preflight the aircraft. With only four passengers and minimal baggage being carried, weight was not a problem, allowing full fuel tanks for a fuel capability of four and a half hours. An en route time of one hour thirty-five minutes was estimated on the instrument flight plan, utilizing a cruise speed of 215 knots. The requested routing followed the airway system south to Kenai at an initial altitude of 10,000 feet, before continuing west across Cook Inlet to their destination of King Salmon. The call sign for the flight was RAM 81.

Weather conditions were expected to be typical for that time of year, with fog, multiple cloud layers and precipitation over much of the southwest region of Alaska. Turbulence and icing were also forecast over portions of the route, but not in any severity that would pose a problem. The aircraft did have anti-icing capability, if necessary.

The scheduled departure time was delayed slightly, as the passengers were running late, but the crew had everyone aboard and engines running within a few minutes after they arrived. Aircraft run-up procedures were normal, with RAM 81 lifting off from runway 05 at Elmendorf at 8:47 local time. An instrument clearance was received before departure, clearing RAM 81 as filed via radar vectors to the Anchorage VORTAC, then on the V436 airway to Kenai VORTAC, and the V456 airway to King Salmon VORTAC, maintaining one zero thousand feet.

Radar contact with Anchorage Departure Control was established almost immediately after takeoff. Departure Control then vectored Ram 81 clear of conflicting traffic in the Anchorage area. After leveling at 10,000 feet the controller switched RAM 81 over to Anchorage Center for further handling at 9:01 am.

---

1 VORTACs are a combination VOR and TACAN navigational aid, providing azimuth and distance indications. VORs are very high frequency omni-directional range stations, while TACANs are tactical air navigation stations with ultra-high frequency range, designed for military applications.
2 Minimum en route altitudes are the lowest authorized altitudes which assure navigational coverage and obstacle clearance between the navigational fixes on that particular airway segment.

Upon initial contact with Anchorage Center the flight was cleared direct to the Kenai VORTAC, then as filed to King Salmon.

The flight was observed by radar approaching Kenai, at which time the controller transmitted the latest altimeter setting. RAM 81 acknowledged and proceeded as previously cleared.

Anchorage Center continued to track the flight on radar past Kenai and on course on the V456 airway until passing TUCKS intersection at 9:25. TUCKS is situated on the west side of Cook Inlet and was the limit of radar coverage in that particular sector. After reaching the intersection, RAM 81 was advised radar service was terminated and to switch frequencies to a different sector controller.

At no time was RAM 81 cleared to a higher altitude, which was necessary to meet the minimum en route requirements for the airway segment between TUCKS and King Salmon. The pilots did not request a clearance to a higher altitude or enter holding procedures at TUCKS until a clearance could be received, and instead continued along the airway toward higher terrain at the plane's last assigned altitude of 10,000 feet.

Radio transmissions recorded by Anchorage Center at the time, indicate RAM 81 attempted contact with the non-radar sector controller shortly after being switched over to that frequency at TUCKS, but radio contact could not be established. The pilot of RAM 81 then switched back to the previous Anchorage Center control frequency at 9:28, advising they were unable to contact the non-radar sector controller and requesting a different frequency. Anchorage Center immediately responded and transmitted an alternate frequency for the other sector controller, which was acknowledged by RAM 81.

A few seconds later RAM 81 attempted contact with the non-radar sector controller again, but because of a high volume of radio traffic the controller did not hear or acknowledge RAM 81's transmission. The pilot then switched frequencies and reestablished contact with the original Center controller at 9:29. Anchorage Center responded to RAM 81 twice, but there was no answer. A that time the controller noticed RAM 81's radar plot to be approximately twenty miles southwest of TUCKS intersection.

Not until thirty minutes later did the non-radar sector controller who was responsible for the airspace between TUCKS and King Salmon, finally inquire about the status with the original Anchorage Center controller. A moment of confusion ensued between the two controllers on the location and status of RAM 81, but they quickly realized the flight continued past TUCKS intersection at 3,000 feet below the minimum en route altitude and had not established radio contact with the non-radar sector controller for that airspace.

Finally realizing the seriousness of the situation, numerous attempts to contact RAM 81 by both controllers and other facilities were initiated. When those failed, additional radio calls were attempted by a commercial airliner operating in the area, with instructions transmitted over the radio in the blind for RAM 81 to immediately climb to 13,000 feet. No answer was received to any of the transmissions, but continued attempts were made until 10:31 that morning.

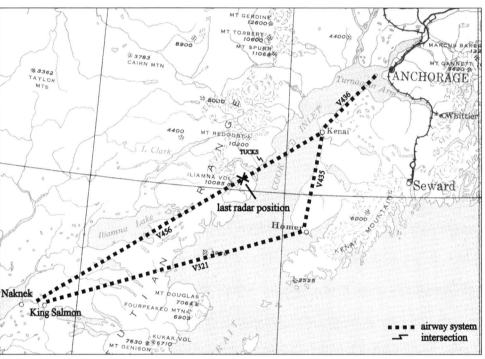

Depiction of the airway system between Anchorage and King Salmon, Alaska.

When RAM 81's estimated time of arrival at King Salmon had passed without any contact being established, an alert notice was sent out to all controllers, tower operators and flight service stations in the region. Once all attempts at locating the flight proved negative, the Regional Control Center (RCC) at Elmendorf AFB and local military commands were notified to begin search and rescue procedures.

Since the missing aircraft's route on the V456 airway was in direct line with Iliamna Volcano, approximately twenty-two miles southwest of TUCKS intersection in the Chigmit Mountains on the western side of Cook Inlet, initial search efforts began concentrating in that area.

Illiamna Volcano is also referred to as Mount Illiamna on some maps and navigation charts. Reaching a height of 10,016 feet, only slightly higher than the flight's last assigned altitude, the volcanic mountain was the highest elevation within twenty-five miles of the airway.

Less than two hours after RAM 81 was reported overdue, an Air Force HC-130 aircraft was in the air and actively searching the area. Two HH3 helicopters, two OV-1 Army Mohawks and a Coast Guard HC-130 from Kodiak joined the effort shortly thereafter, conducting both electronic and visual searches for RAM 81. Infrared photographs were also taken of Iliamna Volcano for possible hot spots indicating where the plane might have crashed.

Reduced visibility and low clouds around the volcano interfered with a thorough search of the area during the afternoon, and by nightfall all the aircraft had returned to their bases without finding any sign of the missing plane. Nothing was observed that provided a single clue regarding RAM 81's whereabouts, but the search crews remained optimistic of eventually finding the aircraft.

U-21 used for pilot training at Ft. Rucker, Alabama. (US Army Aviation Museum-Ref. Collection PN3262)

By the next morning an aircraft flying a weather check of the area reported improved conditions with clear skies around the volcano. Several Army, Air Force and Coast Guard aircraft soon launched in multiple flights that searched the volcano and surrounding mountains. An Army CH-47 helicopter had also been repositioned to Elmendorf AFB and was prepared to launch on a moments notice with a high altitude rescue team if wreckage or survivors were found.

A blackened spot on the side of the volcano was sighted within a few hours that morning by a HC-130, but the potential crash site could not be positively identified until two days later once a low flying helicopter made several passes over the site. What was first thought to be the point of impact where RAM 81 hit the mountain and exploded, instead turned out to be a cluster of exposed rocks that were windswept of snow. No other unusual disturbances were observed on the mountain.

Speculation continued as to the cause of RAM 81's disappearance, with the most likely reason being the failure of air traffic control and the pilots in recognizing an improper minimum flight altitude being maintained after passing TUCKS intersection. Although a possibility of mechanical failure also existed, there was no direct evidence to support that assumption. Flying into extreme turbulence could have caused the missing plane to lose control and crash as well. The phenomenon was not unheard of in Alaska, especially near high mountain areas where weather patterns changed unexpectedly, but the conditions at the time were not conducive to strong in-flight winds and turbulence, and no other aircraft in the vicinity experienced any problems.

Two days after Ram 81 disappeared, the weather again deteriorated, spreading low clouds and precipitation over the mountains and coastal areas of Cook Inlet.

View of Iliamna Volcano (Mt Iliamna) from the east side of Cook Inlet. (Heavenly Sights Charters)

Five military aircraft were still able to cover a narrow sector along RAM 81's proposed flight route that morning, joined by seven other military and civilian planes by early afternoon. Nothing was found.

In Anchorage an internal investigation concerning the involvement of air traffic controllers in the mishap was initiated by the Federal Aviation Administration (FAA). A statement later issued by the Regional Director of the FAA faulted Anchorage air traffic control for not providing a proper altitude clearance for RAM 81, but noted the pilot was also at fault for continuing the flight past TUCKS intersection at an insufficient minimum altitude. One of the most culpable ATC controllers involved was suspended until the investigation was completed.

For the next several days worsening weather hampered and delayed the search efforts, grounding some of the smaller aircraft which were more susceptible to icing and turbulent conditions. Solid clouds from the ground up to 15,000 feet were prevalent over much of the region, making a visual search of the area impossible. On 14 and 15 February search efforts were cancelled in the morning until conditions improved in the afternoon.

A minor break in the weather on the 16th allowed five aircraft to again search the Iliamna Volcano and surrounding mountains, but most of the other aircraft remained on the ground until improved conditions allowed a search of the passes and low areas between Iliamna and King Salmon.

By 17 February the search effort had covered almost 18,000 square miles during the seven day period since RAM 81 vanished. Over the next two days marginal weather conditions over Cook Inlet and the Southwest Peninsula again hampered search activities, but by the 20th they improved considerably, allowing eleven military aircraft to take part. The entire route of the missing plane was covered extensively on either side of the airway system, with no sightings of wreckage or survivors.

Persistent bad weather over the next three days again delayed flights and interfered with the search operation, but every possible area where RAM 81 might have gone down was covered at least once, with many of the more likely areas covered several times. In all, tens of thousands of square miles were over flown on multiple occasions by numerous aircraft. After two weeks of searching and no evidence of Ram 81's location, the operation was officially suspended.

The Army accident investigation and internal FAA investigation continued for several weeks, but nothing conclusive was determined that could directly establish the whereabouts of RAM 81.

Although several facts were uncovered by the Army that could have contributed to the accident, they were not believed to be significant. These included a finding that the oxygen constant flow regulator and pressure reducer aboard the aircraft were ten days overdue for replacement. Analysis of the oxygen servicing system in the hangar, which was used for filling the oxygen containers on the aircraft, was also found below standard. The methane level in the system was higher than the minimum authorized, but was still determined to be in a safe amount that would not have influenced or altered an individual's performance.

A maintenance requirement for the installation of an encoding altimeter on the aircraft had not been complied with, and was of no use on the flight. Anchorage Center did not have the capability to receive automatic altitude information in that sector. A radar altimeter, which provides an indication of the aircraft's actual height above the terrain, was also not installed.

An overdue battery capacity check was found during a review of the aircraft's maintenance records, but battery failure was dismissed as a factor since a required seven day battery inspection was performed two days before the flight.

The TACAN receiver on the aircraft was found to have a history of intermittent DME operation, which when operating correctly would provide an accurate distance indication from selected navigational beacons. Other pilots in the unit considered that particular receiver unreliable and it had been written up in the logbook several days before. Since there was no replacement available in the unit at the time, it was decided to keep the receiver installed, in order to at least provide intermittent DME information to the pilots. Captain Deeter flight checked the functions of the receiver two days before the mission and signed it off as okay.

Normal procedure was for the pilots in the unit to only use TACAN receivers in the aircraft for distance information, especially when fixing airway intersections such as TUCKS. The only other instrument navigational equipment available on the aircraft was a VOR and ADF receiver. Even if the DME portion of the TACAN receiver failed, the Homer ADF could have been used in conjunction with cross-tuning the Homer and Kenai VORs to fix the intersection, assuming the pilots were aware of the malfunction with the TACAN.

A weather radar system normally installed on the aircraft had been recently removed for repair. Although not normally used for such a purpose, the accident in-

U-21 with the cabin door down, awaiting passengers. (US Army Aviation Museum-Ref. Collection PN6022)

vestigation board felt the possibility existed that higher terrain along RAM 81's flight path could have been identified by the pilots if an operational system was installed.

No other discrepancies were found in the aircraft records. The plane was determined to be airworthy and suitable for the mission, with weight and balance computations within allowable limits.

CW4 Battle was considered a highly experienced aviator with almost 6,500 total flight hours, over half in fixed-wing aircraft and more than 500 hours in instrument conditions. Over his career he had accumulated nearly 2,000 hours as a flight instructor and instrument examiner, and was regarded as extremely knowledgeable in instrument procedures. His last flight to King Salmon using the V456 airway route was the previous summer.

Captain Deeter was also considered an excellent pilot. He possessed more than 1,600 total flight hours, including 200 hours in instrument conditions and was rated as a fixed-wing instructor pilot.

From witness statements taken during the investigation it was known that Battle and Deeter discussed the route and the higher minimum en route altitude requirements on the V456 airway during their flight planning. This route was believed to be selected instead of the more southern route used by unpressurized aircraft, because the Homer VORTAC was out of service and unable to provided accurate distance information over the V436 airway to King Salmon.

Prior to the mission the pilots were overheard discussing oxygen requirements for the higher minimum en route altitude on the V456 airway. An initial altitude

U-21 in cruise flight over the Midwest United States. (US Army Aviation Museum-Ref. Collection PN291)

of 10,000 feet was probably filed with the intent of being cleared by ATC to a higher altitude before passing TUCKS intersection. This would have allowed the first forty minutes to be flown without the use of oxygen regulators, before climbing to a higher altitude later in the flight.

Traffic volume being monitored by air traffic control at the time of RAM 81's flight was considered moderate to high, with 128 different contacts visible on air traffic control's radar scope. The Anchorage Center radar controller failed to realize RAM 81 was cleared to an inappropriate altitude for the next route segment beginning at TUCKS intersection, and improperly briefed the relieving controller that the aircraft only required radar termination upon passing the intersection. At the time he was also handling the radar traffic for another sector controller who was on break.

The relief controller failed to notice RAM 81 was cleared at an altitude below the minimum established and did not notify the flight when they were fixed by radar over TUCKS intersection. The call, although not mandatory, was routinely sent by Anchorage Center and would have been expected by the pilots. The relief controller did, however, notify RAM 81 that radar service was terminated and to contact the non-radar controller on a different frequency.

In the non-radar control sector where RAM 81 was transferred to, a controller in training who was at the time under direct supervision, accepted and posted the flight strip for RAM 81. The strip clearly showed an assigned altitude below the minimum required. Neither the supervisor nor trainee noticed the error. Because of their own apparent traffic volume in the non-radar sector, attempts at contact from RAM 81 went unmonitored and its status was forgotten for the next half hour.

One of the air traffic controllers responsible for the handling of RAM 81 admitted during the FAA's internal investigation of receiving only two hours of sleep the previous night.

Even though several of the air traffic controllers were in violation of Federal Aviation Regulations by not clearing RAM 81 to a higher altitude for the next route segment after TUCKS, responsibility was clearly shared by the pilot-in-command. It was the pilot's responsibility to either climb or request a climb to a higher minimum en route altitude when a clearance had not been issued in advance. This was not done and a continuing sequence of errors resulted in the subsequence loss of the aircraft and crew.

Why two highly experienced pilots failed to notice their dangerous predicament will remain a mystery, but the accident board believed the pilots falsely relied on air traffic control to provide their position and terrain avoidance. The reliability of the aircraft's TACAN receiver was determined to be questionable and since air traffic control did not advise them upon reaching TUCKS intersection or clear them to a higher altitude, as would be expected under radar contact, it is possible the pilots were unaware of their location on the airway. Further distractions from repeated attempts at contact with ATC on several frequencies probably interfered with the pilots focusing proper attention to their position on the airway.

The minimum obstruction clearance altitude for the airway between TUCKS and Iliamna was depicted as 12,000 feet on the instrument chart, meaning navigational indications from the Kenai VOR would probably have been erratic or unreliable the further the aircraft flew away from TUCKS intersection at their altitude of only 10,000 feet.³ Weak reception might have caused the pilots to falsely assume they were further east.

From forecasts and weather data analyzed after the accident, the investigation board determined RAM 81 was in instrument conditions from the time it passed the Kenai VORTAC. Cloud layers at the time extended from 5,000 feet up to 12,000 feet, as verified by other aircraft operating in the area. Clear icing was also present and was reported as building rapidly on a commercial aircraft in the same vicinity before it could climb and exit the conditions. RAM 81 gave no indication of encountering icing conditions or other hazards to flight.

Did RAM 81 impact Iliamna Volcano? Winds aloft at 10,000 feet would have reduced RAM 81's groundspeed to 180 knots after turning on course on the

U-21A. (Raytheon Aircraft)

V456 airway, resulting in a time en route of seven minutes between TUCKS intersection and the mountain. Air traffic control plotted the aircraft passing the intersection at 9:22 and received the last radio call from RAM 81 at 9:29.

That by itself does not necessarily mean the aircraft impacted the mountain, since the highest point was only slightly higher than their cruising altitude of 10,000 feet. RAM 81 could have missed the terrain entirely and continued on toward King Salmon. Once the aircraft passed Iliamna Volcano the higher terrain would have been behind their flight path and could have effectively blocked further radio transmissions with Anchorage Center. But why would RAM 81 fail to contact King Salmon as the flight progressed and where would it have gone?

---

3 Minimum Obstruction Clearance Altitude is defined as the lowest altitude providing obstacle clearance requirements for the entire route and acceptable radio reception within 22 nautical miles of a VOR.

Since there is no evidence of what actually transpired, one could speculate the aircraft did in fact climb above 10,000 feet while attempting to contact the non-radar sector controller after passing TUCKS intersection. After all, CW4 Battle was an extremely competent, experienced instrument examiner with extensive knowledge of instrument flight regulations. Unable to establish contact with the controller, he might have decided on his own to climb higher, even without a clearance, because of poor navigation reception which precluded entering a holding pattern at TUCKS. The possibility could then have occurred, although remote, that the aircraft experienced some sort of sudden and catastrophic structural failure which prevented a distress call from being sent.

Even if RAM 81 did impact Iliamna Volcano the likelihood of spotting the wreckage from the air was not a certainty. An avalanche or new snowfall could have covered the wreckage in a short period, completely hiding any sign of a crash from searching planes. The wreckage could have slid or been thrown by the force of impact into one of many narrow cuts or creviced glaciers hundreds of feet deep, easily hiding any evidence of the plane from view. Heavy accumulations of past snowfalls also littered the sides of the mountain, where strong winds often pushed the powder into deep snow drifts capable of hiding the largest aircraft.

Whether or not RAM 81 crashed into the volcano might never be determined with any certainty, but if what remains of the aircraft and its seven occupants lies hidden beneath the snow and ice, they are not alone. In 1958, only three days before Christmas, a USAF C-54G carrying fifteen servicemen hit the eastern slope of the volcano, killing everyone aboard. A second aircraft, a civilian DHC-6 owned by Alaska Aeronautical Industries with thirteen individuals, fatally impacted the southern slope in September 1977. None of the bodies from either disaster has been recovered.

Bibliography

Books

Andersson, Lennart. *Soviet Aircraft and Aviation*. Annapolis, MD: Naval Institute Press, 1994.

Chant, Christopher and Taylor, Michael. *The World's Greatest Aircraft*. Edison, NJ: Chartwell Books, 2003.

Cohen, Stan. *The Forgotten War*. Vols. 1-4. Missoula, MT: Pictorial Histories Publishing Company, 1981-1993.

Gann, Ernest K. *The High and the Mighty*. New York: William Sloan, 1953. *Flying Circus*. New York: Macmillan Publishing, 1974.

Garfield, Brian. *The Thousand Mile War*. New York: Ballantine Books, 1971.

Gero, David. *Military Aviation Disasters*. Somerset, UK: Patrick Stephens, 1999. *Aviation Disasters*. Somerset, U.K. Patrick Stephens, 1994.

Gregory, Glenn. *Never Too Late to Be a Hero*. Seattle: Peanut Butter Publishing, 1995.

Hutchison, Kevin. *World War II in the North Pacific*. Westport, CT: Greenwood Press, 1994.

Jackson, Donald. *The Explorers*. Alexandria, VA: Time-Life Books/The Epic of Flight, 1983.

Kirk, John and Young, Robert. *Great Weapons of World War II*. New York: Bonanza Books, 1961.

Kurilchyk, Walter. *Chasing Ghosts*. Capistrano Beach, CA: Aviation History Publishing, 1997.

Lyon, Thoburn. *Practical Air Navigation*. New York: Eastern Printing, 1948.

McCannon, John. *Red Arctic*. New York: Oxford University Press, 1998.

Mondadori, Arnoldo. *The Illustrated Encyclopedia of Military Aircraft*. Edison, NJ: Chartwell Books, 2001.

Morrison, Robert. *Russia's Shortcut to Fame*. Vancouver, WA: Morrison and Family Publishing, 1987.

Nichols, Reeder. *Pilot's Radio Manual*. Washington: GPO, 1940.

Smith, Blake. *Warplanes to Alaska*. Blain, WA: Hancock House Publishers, 1998.

Newspapers and Magazines
Anchorage Daily News. Multiple articles. Anchorage, AK. 1972, 1992, 2000.
Anchorage Daily Times. Multiple articles. Anchorage, AK. 1945, 1947-1948, 1950-1952, 1969, 1972.
Anchorage Times. Multiple articles. Anchorage, AK. 1977, 1987-1988.
Borutski, Barry. "Cold War Relic." *Flight Journal.* Dec 2000.
"Cargo Plane Missing in North." *Ketchikan Alaska Chronicle.* Ketchikan. Jul 1943.
Daily Alaska Empire. Multiple Articles. Juneau, AK. 1945, 1948, 1950.
Fairbanks Daily News Miner. Multiple articles. Fairbanks, AK. 1936-1938, 1943, 1945, 1947-1948, 1950-1952, 1969, 1977, 1987, 1999, 2003.
Ford, Daniel. "B-36: Bomber at the Crossroads." *Air&Space.* May 1996.
Jonz, Don. "Ice Without Fear." *Flying.* Nov 1972.
Juneau Empire. Multiple articles. Juneau, AK. 1987.
Kellems, Homer. "We Tried to Solve an Arctic Mystery." *The Alaska Sportsman,* Jun 1940.
Kodiak Mirror. Multiple articles. Kodiak, AK. 1947-1948.
Moreth, Ed. "1943 Crash Site Rediscovered." *Alaska Magazine.* Mar 1988.
"Army IDs crash victim." *Commandant's Bulletin.* Jul 1992.
"Homeward Bound." *Alaska Bear.* Oct-Dec 1987.
Morrisette, Stephen. "Whatever Happened to Capt Ernie Walker?" *Flying Safety.* Mar 1988.
New York Times. Multiple articles. New York. 1937-1938.
Seattle Post-Intelligencer. Multiple articles. Seattle. 1947, 2002.
Septer, Dirk. "Broken Arrow." *Airforce.* Winter 1998.
Stroganov, Oleg. "Polar Tragedy." *Soviet Life.* Sep 1988.
Wetterhahn, Ralph. "One Down in Kamchatka." *Retired Officer.* Jan 2001.
Whitehorse Star. Multiple articles. Whitehorse, Yukon Territory. 1950-1951.
Wilkins, Hubert. "Our Search for the Lost Aviators." *National Geographic Magazine,* Aug 1938.

Government Publications and Documents
Canada. Department of Transport. *Summary Accident Report.* 1951.
United States. Army Air Forces. *Instrument Flying Advanced.* GPO, 1944.
Army Air Forces. *Missing Air Crew Report.* 1946.
Army Air Forces. *Report of Major Accident.* 1947, 1950.
Army Air Forces. *Report of Major Accident Supplemental Report.* 1948.
Civil Aeronautics Board. *Accident Investigation Report.* 1949.
Department of the Air Force. "Extract of Elmendorf Aeronautical Station Tapes." 1969.
Department of the Air Force. "Final Mission Report." 1972.
Department of the Air Force. "FTD Analysis Transcript." 1969.
Department of the Air Force. "Narrative Report on Operation Mike." 1950.
Department of the Air Force. *Report of AF Aircraft Accident.* 1952-1953.
Department of the Air Force. "Search and Rescue Operations Report." 1969.
Department of the Air Force. *Theory of Instrument Flying.* GPO, 1951.
Department of the Air Force. *USAF Accident/Incident Report.* 1969.
Department of the Army. *Aircraft Accident/Incident Report.* 1977.
U.S. Department of Transportation. *FAR/AIM.* Aviation

Supplies and Academics, 2002.
National Oceanic and Atmospheric Administration. *Sectional Aeronautical Chart.* 1985-2005.
National Transportation Safety Board. *Aircraft Accident Report.* 1973.
Navy Department. "Administrative Report." 1946, 1948.
Navy Department. *Aircraft Accident Report.* 1947-1948.
Navy Department. *Aircraft Action Report.* 1944, 1946-1948.
Navy Department. *Aircraft Trouble Report.* 1944.
United States Pacific Fleet. "Fleet Air Wing Four History." 1945.
U.S. Army Air Forces, *Report of Aircraft Accident.* 1943-1944.
U.S. Army Central Identification Laboratory. "Search and Recovery Report." 1987, 2001.
U.S. Coast and Geodetic Survey. *World Aeronautical Chart.* 1943-1967.

Video Cassette
Last Flight of Bomber 31. Videocassette, NOVA, 2003. 60 min.

Internet
Ace Pilots. 16 Jan, 2003 http://www.acepilots.com/planes/b29.html
Aerofiles. 18 Nov, 2003 http://www.aerofiles.com/_beech.html
Aerospace. 2002 http://www.aerospaceweb.org/
Air Force Museum. 2003 http://www.afmuseum.com/aircraft/
Airliners. 9 Feb, 2004 http://www.airliners.net/
Aviation Enthusiast Corner. 2003-2204 http://www.aero-web.org/
Aviation History. 2003 http://www.aviationhistory.com/
Boeing. 2003 http://www.boeing.com/companyoffices/history/
Bolkhovitinov DB-A. 15 Dec, 2002 http://www. alpha1.fsb.hr/~ah951096/avi/bolkh.html
Centercomp. 29 Nov, 2004 http://www.centercomp.com/cgi-bin/dc3/stories
Check-six. 2002-2005 http://check-six.com/
Clearwater, John. "Peace BC-1950 Bomber Crash in BC." Alberni Environmental Coalition. 9 Oct, 2002 http://www.portaec.net/library/
CNN. 4 Feb, 2002 http://cnn.com/
CSD. 2003 http://www.csd.uwo.ca/~pettypi/elevon/baugher_us/
Daves Warbirds. 25 Mar, 2003 http://www.daveswarbirds.com/
Davidge, Doug. "Environmental Impact Study of Crash Site of USAF Bomber." Cowtown. 9 Oct, 2002 http://www.cowtown.net/
Defenselink. 28 Nov, 2003 http://www.defenselink.mil/
Douglas DC3. 8 April, 2003 http://www.douglasdc3.com/dc3specs/
Earthpulse. 2001-2002 http://www.earthpulse.com/
FAS. 16 Jan, 2003 http://www.fas.org/nuke/guide/usa/bomber/b-29.htm
Goleta Air and Space Museum. 9 Oct, 2002 http://www.air-and-space.com/
Jackal Squadron. 14 May, 2003 http://www.jackalsquadron.org/pages/content/articles/ pv1_cd/
Kingdon Aviation. 6 Nov, 2003 http://www.kinghawes.com/
Kiwi aircraft. 12 Jan, 2004 http://www.kiwiaircraftimages.com/kingair.html
Landings. 2 Nov, 2001. http://www.landings.com/

Maslov, Mikhail. "Aviation-Time." *Aviation Magazine.* 2 Jun, 2004 http://www.aviation-time.kiev.ua/eng/article.php?IDA=41

MILNET. 4 Feb, 2002 http://www.milnet.com/

Mostyn, Richard. "Broken Arrow." *Yukon News.* 9 Oct, 2002 http:// www.yukonweb.com/

Navylib. http://www.navylib.com/Specs.htm

Perkins, Frank. "Nightmare at Mignight." *Cowtown.* 9 Oct, 2002 http://www.cowtown.net/

"Rivet Amber." *Williwaw.* 25 Nov, 2001 http://www.hlswilliwaw.com/

Robison, Tom. "The B-29 in Weather Reconnaissance." 5 Mar, 2003 http://home.att.net/~sallyann4/robison-col.html

Savine, Alexander. *V.F. Bolkhovitinov.* 15 Dec,2002 http://www.ctrl-c.liu.se/misc/ram/db-a.html

7th Wing Operations. 9 Oct, 2002 http://www.7bwb-36assn.org/

Turyshev, Slava. "The Lost 1937 Arctic Flight of Soviet Aviation Hero Sigismund Levanevsky." *Los Alamos National Laboratory.* 14 Dec, 2002 http://www.lanl.gov/physics/colloquium/abstracts

USCG. 2002 http://www.uscg.mil/

US Navy. 2002 http://www.history.navy.mil/

Warbird Alley. 2002-2003 http://www.warbirdalley.com/

Wright/Patterson AFB. 2003 http://www.wpafb.af.mil/museum/

Yeletsky, Viktor. "The Search for Levanevsky." *NWTANDY.* 2 Feb, 2004 http://www.nwtandy.rcsigs.ca/levanevsky.htm

Interviews

Acord, Randy. Ladd Field/Cold Weather test pilot, Pioneer Aviation Museum curator. Fairbanks, AK.

Behn, Georgie. Hunting guide who found C-48 wreckage. Fort Nelson, BC.

Downing, Janet. Former wife of Pan Alaska Airways chief pilot. Fairbanks.

Malone, Tom. Old Crow site investigator. Fairbanks.

Millard, Doug. Former Wien Airlines captain. Wasilla, AK.

Mills, Carl. Aviation historian and author. North York, ON.

Norton, Dave. Administrative Liason for Old Crow site investigation. Fairbanks.

Pederson, Sverre. Project leader for Old Crow site investigation. Fairbanks.

Stone, David. Scientist at the University of Alaska Geophysical Institute. Fairbanks.

Papers

Gotthardt, R.M. "Report on the 1989 Levanevsky Search Expedition, Sam Lake, Northern Yukon." Yukon Science Institute, 2002.

Owens, Melissa. "The Rivet Amber." 1995.

Index

10th Rescue Squadron ... 110, 112, 133, 143-145
11th Air Force ... 63
196th Regimental Combat Team ... 206
222nd Aviation Battalion ... 243, 245
28th Bomb Group ... 107, 110, 112
436th Bomb Squadron ... 169, 171
6th Strategic Reconnaissance Wing ... 221
8th Air Force ... 153, 173
A-20, Douglas ... 56
Adak ... 63, 109-110, 121, 125, 133, 135, 138, 212, 223
ADF, Automatic Direction Finder ... 252
Adkins, Orval ... 64
Air National Guard ... 154
Air Transport Command ... 37, 40-41, 43, 54, 160
Air Transport Wing ... 44
Aishihik ... 55, 60, 157, 165
Aklavik ... 22-23, 26-31
Alaska Aeronautical Industries ... 256
Alaska Air Command ... 107-108, 154
Alaska Coastal Airlines ... 195
Aleutian Islands ... 63-64, 77, 117, 127, 131, 135, 189, 217, 227
Aleutian Range ... 129
Aleutians ... 53, 63-65, 67, 73-74, 81, 108, 110, 118, 122, 125, 130, 132, 139, 144, 212, 219
Amber 1 or Amber One, airway ... 98
Amber 2 ... airway, 39, 41, 48, 155, 157
Amber One ... 203
Amchitka ... 63-64, 219
Amderma ... 20
Anadyr River ... 12

Anchorage 64, 66, 69, 105, 107, 113-114, 124,
142-143, 146, 155-157, 172-174, 189-190, 192, 194, 198, 206-207, 209-210, 212,
218, 230, 232, 234-235, 237-238, 243, 246-248, 250-251, 253-255, 258
Annette 102, 143, 145-146, 173-174
ANT-25, Tupolev 11
Arcano, Douglas 228
Archangel 20, 26
Arco Alaska 34-35
Arctic Institute of North America 34
Arctic Ocean 12, 19, 25-26
Armistead
Armistead, Clyde 13-15
Ascol, Holiel
Ashdown Island 180
Askildson, Lloyd 130, 138
Atomic bomb 170
Atomic Energy Commission 188
Attu 63, 65, 73-79, 81-82, 87, 93
Atwood, Charles 40
B-17, Boeing 124, 126, 133, 192
B-18A, Douglas 64
B-24, Consolidated 93-94, 129-130
B-25H, North American 64, 66-67, 69-71
B-29, Boeing 105-115, 259-260
B-36B, Convair 163, 170-172, 174-176, 178-179, 181-183, 185-187
Baidukov, General 34
Baker, Charles 95
Bangert, Pervis 121
Banks Island 176
Baranof Island 100, 102, 190
Barden
Barden, Franklin 130
Barents Sea 20
Barker, Paul 130-132, 135, 138-140
Barrett, William 244
Barrow 22-26, 28-32, 245
Barry, Harold 5, 172-177, 180-181, 258
Barter Island 22-23, 26-28, 31-32, 35
Battle, Ralph 86, 106, 120, 169, 243-245, 252-253, 256
Beasley, Horace 228
Beaton River 39, 41
Beaufort Sea 24

Index

Beaver Creek 157
Beech Point 32
Begich, Nicholas 229-230, 232-233, 236
Benevides, Eugene 228
Bering Sea 56, 64, 105, 108, 112, 115, 129-130, 133, 135, 217, 223-224, 227-228
Bering Strait 12-13, 22, 53
Bethel 113-114, 217
Big Delta 55, 58-59
Biggs Air Force Base 153
Bittersweet, Coast Guard ship 133
Blakeman, Leon 107
Boggs, Hale 229-233, 236, 239-240
Bolkhovitinov, Victor 11, 17, 259-260
Boon, Arthur 199
Boyd, Edwin 209
Brittain, Gerald 154
Broad Pass 68-69
Broken Arrow 7, 169, 178-179, 186, 188, 258, 260
Brooks Range 22, 30, 33
Brown, Russel 6, 228, 230, 233
Bruin Pass 105, 107-108, 110-112
Bumbus, Herman 95
Burwash Landing 157, 163
C-119C, Fairchild 205-206, 208-209, 212-215
C-124A, Douglas 212
C-130, Lockheed 219, 235
C-45F, Beech 102
C-48B, Douglas 58-60
C-49K, Douglas 37, 39-40, 43, 46
C-54D, Douglas 153-155, 160-161, 164, 166
C-54G, Douglas 256
CAA 147-148
CAB 146-148
Calgary 155, 162
Canadian Department of Transport 196-199
Canadian Pacific Airlines 5, 189-192, 194, 197-198, 200
Canning River 22
Cape Lopatka 74, 81-82
Cape Schmidt 21-22
Cape Spencer Intersection 142-144, 148, 150-151, 191, 193-195, 200, 203
Cape Yakataga 64

Carcross ... 161, 163
Carpenter, Paul ... 228
Carswell Air Force Base ... 169
Casco Field ... 73, 79, 93
Cayuga, Canadian Navy ship ... 179
Central Identification Laboratory ... 5, 83, 85, 89, 259
Cheliuskin, Soviet ship ... 12-15
Chichagof Island ... 102
Chigmit Mountains ... 66, 112, 248
Christy, Lee ... 64
Civil Aeronautics Administration ... 143, 147, 200
Civil Aeronautics Board ... 143, 146, 258
Civil Air Patrol ... 194, 232, 234, 241
Clark, William ... 130, 138
Clinton, William ... 240
Cocopa, US Navy ship ... 179
Cold Bay ... 118-119, 123-124, 126, 129, 133, 135-136, 138, 140
Cold War ... 33, 83, 105-106, 179, 220, 228, 258
Cold Weather Testing Unit ... 64
Coleman, William ... 130, 138
Colville River ... 30
Consolver, Sherman Jr. ... 228
Cook Inlet ... 66, 107-110, 112, 115, 156, 205, 207-209, 230, 245-248, 250-251
Coppermine ... 23-24, 26, 31
Cordova ... 190, 192, 230, 237
Cox, Ernest
Crosson, Joe ... 22, 28
Crown, Samual ... 80
D-18S, Beechcraft ... 102
Darrah Roy ... 173
DBA Bolkhovitinov ... 11, 259
DC-3A Douglas ... 58, 64
DC-4 Douglas ... 5, 50, 153, 167, 189-191, 193-200-203
Deeter Donald ... 243, 244, 252, 253
Delta Island ... 205-207
Department of National Defense ... 187
DEWIZ ... 219
DHC-6 DeHavilland ... 256
Dickson Island ... 20
Direction Finding station ... 110, 114
Distant Early Warning Identification Zone ... 219
DME ... 251-252

Dreber, Charles 228
DST-A, Douglas 58
Dutch Harbor 117, 119-120, 122-124, 126
Duval, John 121
Edge, John 7, 20, 64, 91, 117-118, 122, 127, 141, 176, 229, 244
Edmonton 18, 22-23, 29, 37, 41, 44, 49, 54, 56, 59, 159, 162, 194
Eichorn, Robert 130, 138
Eielson Air Force Base..169, 217
Eighth Air Force..170
Electra, Lockheed 22-24, 28-30
Elmendorf Air Force Base 66, 105, 133, 153, 189-190, 194, 205, 232, 243
ELT..233, 237, 240
Emergency Locator Transmitter...233
Empire Express 7, 73
Enger, Gill 54-59, 61-62
Environmental Protection Branch..187
FAA.. 241, 250-251, 254
Fairbanks...11, 13-14, 16, 18-25, 28, 40-41, 53-56, 58,
64, 66, 68-69, 71, 106, 169, 192, 212, 217, 221, 233, 258, 260
Fairchild 71 22
Fairchild 82 23
Fairweather, Mountain Range...150, 193
Fat Man, MKIV atomic bomb...170, 176, 178, 186
Federal Aviation Administration 250
Fleet Air Wing Four 63, 73, 80, 122, 130, 259
Flight Service Station...230
Ford 4AT 23
Ford, James 23-24, 173, 258
Fort Greely 243
Fort Nelson...37-45, 47-48, 50, 159-160, 162, 260
Fort Richardson...243
Fort St. John 37, 39-43, 48-49, 155
Fort Wainwright 243
Fort Yukon..69
Fox, Robert 228
Fox, Victor 198
Franz Josef Land 20, 22, 28
Freedom of Information Act 181, 186, 238
Frentz, Kenneth 107
Fridley, Clarence 80
FSS..230, 235
Ft. Glenn 118-119, 122, 124

265

Ft. Randall 109-110, 119, 121-123, 125
Ft. Richardson 124-125, 206
Fuleihan, Nave 121, 127
Galkovsky, Nicolai 16
General Airways..167, 193, 201, 203
Geological Survey of Canada 186
Gerhart, Paul 173
Gibson Girl, survival radio 42, 111, 139, 162
Gibson, Clarence..154
Glennallen 69, 238
Godovikov, Nikolai..16
Grand Prairie 37
Grazyansky, Russian pilot...28
Great Falls 41, 47, 53, 57, 155, 158, 160, 162, 180
Green 8 airway 155
Greenland Sea 31
Gregory, Herbert 2, 228, 257
Gulf of Alaska 64, 91-92, 96, 98, 100, 145, 151, 173-174, 189-191, 199, 201, 212
Gulkana 50, 155-156
Gustavus 142, 148, 200
Ham radio 162, 238
Hanlon, John 80
Hebert, Harvey 228
Hecate Strait..180
Hemlock, Navy ship 101
Herndon, Bryce 121
Herndon, Harold 130,138,, 140
Hiroshima..107
Holland, Norman..130, 138
Hollick-Kenyon, Herbert 23, 29-30
Hoover, J. Edgar 239
Iliamna Volcano 248-251, 255-256
Irkutsk 22
Jarvis, Hubert 107
Johnson, William...107, 232, 240
Johnston, Lieutenant 96, 100
Joint Commission.. 83
Jones, Paul...244
Jonz, Don 229-230, 233, 235-237, 240-241, 258
Juneau 144, 167, 192, 194-195, 223, 230-232, 234, 237-238, 258
Kachemak Bay 210
Kamchatka 74, 77, 80-82, 84-85, 87, 105, 221, 258

Kara Sea 20
Kastanayev, Nikolai 16
Kellems, Homer 32, 34, 258
Kenai 66, 109, 142, 194, 206-208, 213, 246-247, 252, 254-255
Kennedy, John 231-232, 239
Ketchikan 173, 184, 195, 234, 258
KGB 83, 89
King Salmon 243-248, 251, 253, 255
King, Maurice 6, 193, 218, 220, 224-228
Kinnear, Andrew 141, 146-147
Kiska 63-64
Kispiox Mountain 184-186
Kluane Lake 165
Kodiak 6, 64, 66, 68-69, 91-92, 94, 96-98, 100-101, 109-110,
117-126, 129-136, 138-139, 192, 206-210, 212, 223, 234, 249, 258
Korean War 189
Krassin, Soviet ship 22, 24, 26-28
Krausher, E.L. 199
Kurilchyk, Walter 34, 257
Kurile Islands 63, 65, 73-74, 77, 82
Ladd Field 5, 40, 54-59, 64, 66, 69, 106, 109-110, 112, 212, 260
Landis, John 209
Lavery, Bill 13, 15
Lee, Eva 64, 197, 199
LeMay, Curtis 181
Lend-Lease 41, 53, 56, 57, 152, 167
Levanevsky, Sigismund 11-25, 27-36, 260
Levchenko, Victor 14-16
Lewallen, Donald 80
Lindbergh, Charles 11-12, 16, 33
Lindsey, Roy 228
Lomax, James 107
Loran, long-range navigation 200
Ludacka, John 107
MacLean, Lieutenant 95-96, 99-101
Maloney, James 107
Marburg, William 64
Marks, Howard 209
Masset 98-100
MATS 212
Mattern 12-13, 22, 24-25
Mattern, Jimmy 98-100

Mavriki, Slepnyov 13
McChord, Air Force Base 182, 192, 212
McConegley, Harry 154
McDonald, Daniel 173
McGovern, George 240
McKenzie River 23
McLaughlin, Robert 107
McMichael, Kyle 153
McMillan, John 91-92, 94-96, 98-102
Meissner, Richard Jr. 228
Mensing Anthony 41-43, 50
Metzler, Joseph 154
Michaud, Charles 228
Middle Cape 118
Middleton Island 197, 212
Military Air Transport Service 189, 212
Mills, Michael 5, 228, 260
Moran, Kathleen 199
Morrison, Robert 257
Morse code 110, 141, 148, 200-201, 226
Mortel, Richard 228
Moscow 11, 15-16, 18-22, 82
Mount La Perouse 150
Mountain Village 112, 114
Mt. Crillon 167, 193, 201, 203
Mt. Gannett 212
Mt. Hubbard 193
Mt. Iliamna 209, 212, 250
Mt. McKinley 69
Murray, Stuart 22
Musgrove, William 130, 138
Mutnovsky Volcano 84, 86
Nagasaki 107, 170
Naknek 108, 111-112, 115, 133, 136, 139-140
Nelson Lagoon 131
Nelson, James 224
Nixon, Richard 239
Nome 12-13
Norman, Benjamin 95, 130, 138
North Cape 12-13
North Pole 7, 11-12, 15-16, 18, 20-22, 25, 30
Northway 55, 59-60, 155-157, 165

Index

Northwest Airlines 40, 42, 50-51, 54, 140, 147, 196, 199-201
Northwest Staging Route 5, 37-38, 41-43, 53, 61
Northwestern Air Command 44
Notice to Airmen (NOTAM) 140
Nunivak Island 108, 217
Old Crow 34, 260
Oliktok Point 32, 35
Order of the Hero of the Soviet Union 14
Orion, Lockheed 32
P-39, Bell 56, 69
P-51, North American 69
P2V-2, Lockheed 134
Pacific Alaska Air Express 134, 142-143, 145-148, 150-151, 203
Pacific Alaska Airways 13, 22
Palko, James 80
Pan Alaska Airways 229, 233, 240, 260
Pan American 64, 197
Paramushiro 74
Parlier, Jack 80
PB4Y-2, Consolidated 92-93, 95-103, 129-133, 135-140, 150
PBY-1, Consolidated 23, 28
PBY-5A, Consolidated 92, 117, 119-123
Peck, Russell 209
Petropavlovsk 84, 86
Phillips, William 173, 180
Plevelich, John 95
Pollard, Elbert 173, 180
Pooler, Charles 173
Port Heiden 129-130
Port Moller 131, 135-136, 138
Portage Pass 230, 232, 236-237
Pratt and Whitney 58, 77, 80, 121, 129, 147, 154, 171, 189, 215, 244
Pribilof Islands 112, 126, 130
Prince of Wales 101, 146, 195
Prince Patrick 24, 30
Prince William Sound 190, 236, 238
Princess Royal 177, 179-180, 183
Probezhimov, Victor 16
Provincial Police 45
PV-1, Lockheed 73-75, 77-82, 84, 87-88, 259
Queen Charlotte Islands 98-100
Randall, Bob 22-26, 109-110, 119, 121-123, 125

Ransom, Orville 107
Rarick, Melvin 64
RC-135E, Boeing 217, 219-221, 223-226
RCAF 44-45, 159, 167, 183, 211
RCC 248
Red Airway 40, 206
Redus, John 57, 61
Regional Control Center 248
Riddle, Paul 107
Rivet Amber 217-218, 220-221, 223-228, 260
Robbins, S.E. 22
Robertson, Jimmie 209
Rogers, Will 6, 32
Rominiecki, Lucian 228
Roy, James 173, 228
Royal Canadian Air Force 44, 100, 179, 197
Royal Canadian Navy 189
Rudolf Island 20, 22, 26, 28, 31
S-43A, Sikorsky 28
SAC 153, 170, 181
Sand Point 118
Sandspit 197
Scherier, Theodore 173, 177, 180, 183-184
Schilsky, Sam 40
Schuler, Richard 173
Scotch Cap 119
Seward Glacier 193
Sharp, Rayv
Shatz, Lester 228
Shelkova, military airfield 18-20
Shemya 61, 63-64, 98, 110, 112, 114, 189, 192, 198, 217-219, 221-222, 224-228
Shimushu 74, 82
Siberian Sea 19
Sitka 96, 98-103, 134, 142-146, 148-151, 189-191, 200-202, 230, 235
Snag 55, 59-60, 157, 160-162, 164-165
Snow, Raymond 154
Somers, Joseph 130, 138
Somerset Island 27
Soviet Embassy 34
Soviet Lindbergh 11, 16, 33
Spengler, John 95
Spy Island 34

SR-71, Lockheed	235-236
St Marys	112
St. Elias Mountain Range	165
St. George Island	126
St. Lawrence Island	112
St. Mathews Island	112
St. Paul Island	129
Staley, Neal	173
Stalin, Joseph	11-15, 36
Steen, Richard Jr.	228
Stefansson, Vilhjalmur	30
Stephens, Martin	173, 257
Storey, US Coast Guard ship	195
Strategic Air Command	153, 170, 173, 181, 183
Streitmahn, Clyde	154
Summit Field	69
Survival radio	42, 139, 162
Swiney, Melvin	244
Tanacross	55, 59
TB-3, Tupolev	17
Thomson, Bruce	198
Thrasher, Dick	173
Tisik, Mike	153
Tokyo Rose	89
Trans World Airlines	39
Trenton, Robert	130, 138
Trippodi, Vitale	173
Truhe, Warren	107
Tu-4, Tupolev	107
Tuchodi Lakes	44, 47
Tupper, Fred	199
Turnagain Arm	114, 230, 236
Turner, Clarence	64
Tutton, Raymond	107
U-21A, Beech	244-245, 255
U.S. Army Air Forces	34, 37, 39-40, 54, 58, 102, 147, 197, 259
Umnak Island	108, 122
Unalaska Island	125-126, 139
Unimak Island	119, 122, 124, 127, 133, 136
Unimak Pass	122, 124
United States Air Force	44, 171
USAAF	77

USAF ..44-45, 49-50, 155, 158-160, 163, 167, 174, 183-185, 209-214, 225, 228, 235-236, 256, 258-259
V-317, airway 230
V-321, airway 245-246
V456, airway 246-248, 253, 255
Vancouver 135, 189-190, 198-200, 257
Ventura, Lockheed 75, 77-78, 80, 84, 89
VFR 108, 129, 230, 233, 241
Visual flight rules 55, 230
VP-20, Patrol Squadron 130
VPB-120, Patrol Bomb Squadron 95
VPB-122, Patrol Bomb Squadron 135, 139
VPB-139, Patrol Bomb Squadron 73, 84
Vultee, V-1A Special 14-17
Wainwright, Jonathan 243
Walker, Ernie..258
Waller, Haywood..95
Warm Wind 7, 205, 208-209
Warren Commission 231-232, 239
Watson Lake... 37, 41-42, 48, 159-160
Weeks Field..20
Whitehorse 37, 41, 54-57, 59-61, 155-156, 158-161, 163, 194-195, 209, 258
Whitfield, Raymond 173
Whitman, Walter 80-82, 86-87
Wiley, Post 32
Wilkins, Hubert 23-31, 258
Williwaw 131, 260
Wilson, Homer 64
Wilson, Richard 141, 146
Wonders, Donald..228
Wooley, James...130, 138
World War II 33-34, 41, 63, 73, 83, 120, 126, 129, 140, 257
World War Two 39-40, 54, 62-63, 66, 94, 105-106, 147, 162, 169, 171, 197, 203
Wrangell Mountain Range 50
Wright, engine manufacturer 40, 58, 66, 107, 115, 260
Wyonna, US Coast Guard ship 179
Yakobi Island 195
Yakutat 134, 141-142, 145-146, 148-151, 167, 174, 190-196, 200-203, 230, 232, 238
Yakutsk..21
Zhukov, Russian pilot 24, 26-28
Ziegler, William 121-123, 127